S0-BRB-894

Heat treatment of welded steel structures

Heat treatment of welded steel structures

D N Croft

ABINGTON PUBLISHING
Woodhead Publishing Ltd in association with The Welding Institute
Cambridge England

Published by Abington Publishing
Abington Hall, Abington
Cambridge CB1 6AH, England

First published 1996

British Library Cataloguing in Publication Data
A catalogue record for this book is available from the British Library.

ISBN 1 85573 016 2

Cover design by The ColourStudio, cover photograph reproduced courtesy of Cooperheat (UK) Ltd.
Typeset by Best-set Typesetter Ltd, Hong Kong.
Printed by St Edmundsbury Press, Bury St Edmunds, Suffolk, England.

Contents

Preface

Welding is perhaps one of the most common industrial manufacturing processes in use today, as is evidenced in all walks of life. Many materials are joined by the process, both metals and non-metals, to create a wide variety of products. The most common perception of welding relates to joining of metals, and incorporated in metal welding technology is a range of heat treatment applications which enhance the process, given certain conditions, and are used to ensure success of the fused joints in performing to the requirements of the designer.

This book is essentially a revision of the original *Heat Treatment of Welded Structures* by Professor F M Burdekin, addressing developments in application and references to code requirements pertinent at the time of writing. The essential concept has been retained, albeit relating only to those aspects associated with steels in view of their predominance as an engineering material.

Acknowledgements

The author is grateful to Cooperheat (UK) Limited, Mitsui Babcock Energy Limited and TWI for their kind co-operation in providing illustrative material for this book, and to Professor F M Burdekin of UMIST and Dr P Woollin of TWI for their much appreciated advice during preparation of the text.

Extracts from British Standards are reproduced with the permission of BSI. Complete editions of the standards can be obtained by post from BSI Customer Services, 389 Chiswick High Road, London W4 4AL, UK. The American Society of Mechanical Engineers can be contacted at 345 East 47th Street, New York, NY 10017-2392, USA.

Introduction

Welding is one of the most common processes used in today's manufacturing and engineering industries. Many materials are joined by welding, non-metals and metals, to create a wide range of structures from domestic consumer goods to highly complex heavy industrial installations. Perhaps the most common perception of the welding process relates to joining of metals, and the most common of all the metals used is steel in its many forms. These range from basic low-carbon ferritic structural steels, increasing in complexity with the addition of alloying elements to produce grades with applications in load bearing and pressure part manufacture, and on to austenitic and stainless steels which in turn encompass a vast array of alloy types.

A whole technology has evolved around the development of welding processes, welding consumables and weldable alloys, and in particular their application to joining steels in manufacturing engineered products. However, it is often the case that joining of components by the fusion process alone, that is welding, results in a joint incapable of meeting the requirements of the structural design which is customarily based on the properties of the parent material.

To increase the range of parent metals which can be welded, and to ensure the success of the welding process and the integrity of the final product, additional heat is applied in a scientific and controlled manner. Therefore, intrinsic to welding technology is a branch of thermal engineering which produces a series of defined heat treatments contributing towards these aims whilst ensuring the fundamental material properties.

Whilst recognising that the principles of such heat treatments are applicable to many situations which involve joining of metals by welding, in the following chapters the author sets out to introduce their application to steel in view of its predominance as an engineering medium.

Chapter 1

The welding process and its effects

In the majority of cases, the welding process applied to metals joins two components together by fusion. The surfaces to be joined are raised locally to melting point by a source of heat provided by a variety of welding methods based on an electric arc, electrical resistance or a flame. Small or thin components may be fused together autogenously to form the joint, but often a filler material is introduced. The process energy creates a localised molten pool into which the consumable is fed, fusing with the component surface and/or previously deposited weld metal.

As the molten pool is moved along the joint axis, the components are heated, non-uniformly and subsequently cooled, also non-uniformly. Neighbouring elements of material try to expand and contract by differing amounts in accordance with the sequence of the localised thermal cycle.

An analogy can be drawn with casting of metals, except that the welding process is extremely localised and semi-continuous. Characteristically the cooling weld metal zone contracts under conditions of severe restraint, leading to introduction of thermally induced stresses.

As contraction tries to take place and the stress system strives to reach its lowest level to achieve stability, distortion occurs as yielding takes place, see Fig. 1.1.

High stresses coupled with joint restraint and low ductility may lead to failure in the form of cracking. The origins of the residual stresses induced by the thermal cycles of the welding process can

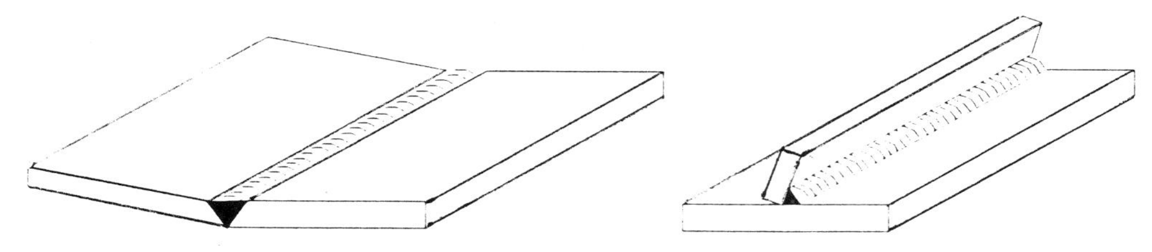

1.1 Distortion caused by welding.

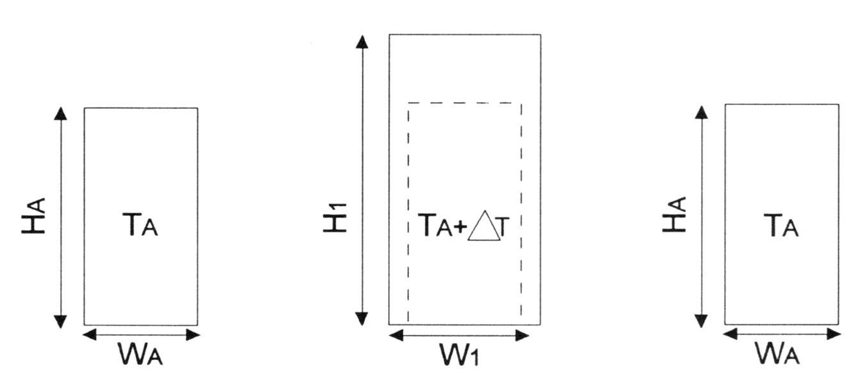

1.2 Unrestrained expansion and contraction.

be explained by considering a 2-dimensional rectangular body, see Fig. 1.2.

Metals expand when heated and, if unrestrained, change their dimensions unilaterally according to their coefficient of thermal expansion. When the body is heated from ambient temperature T_A by an amount ΔT, then the dimensions H_A and W_A increase to H_1 and W_1 respectively, given by

$$H_1 = H_A (1 + \alpha \Delta T)$$
$$W_1 = W_A (1 + \alpha \Delta T)$$

Where α is the coefficient of thermal expansion, typically $12.7–14.6 \times 10^{-6}\ K^{-1}$ for ferritic steels.

On cooling to ambient, the body will revert to its original shape. Now place a restriction on free expansion in the vertical axis, the H dimension in Fig. 1.3.

If the temperature is raised as before, the body cannot increase

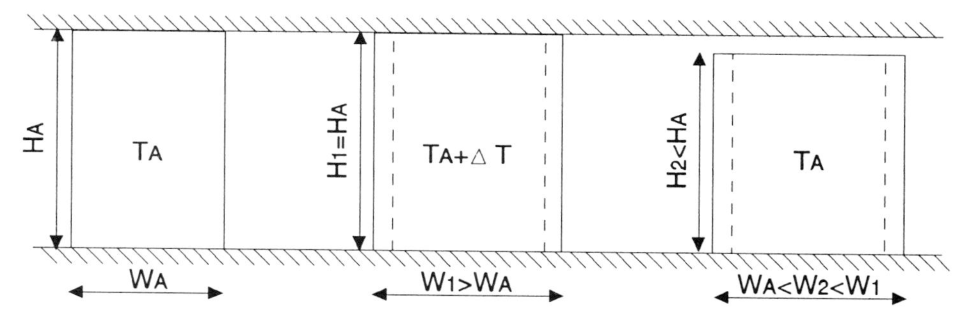

1.3 Restrained expansion, unrestrained contraction.

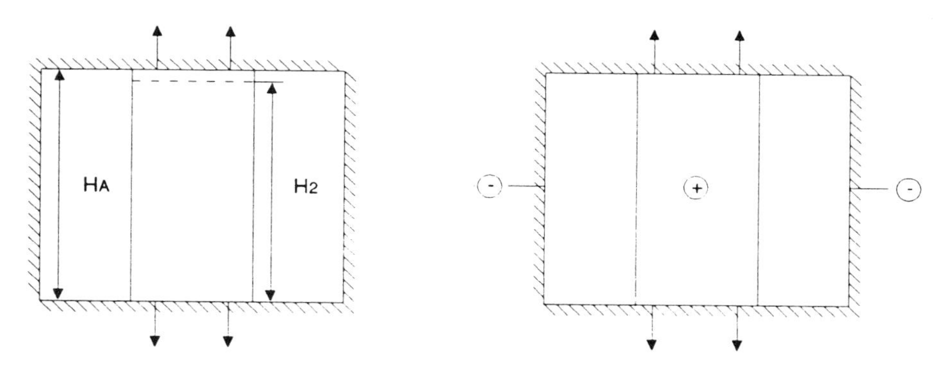

1.4 Restrained expansion and contraction.

its dimension in the H direction. The only change that can occur is in the W dimension, which undergoes a larger expansion than in Fig. 1.2. If the change in temperature is sufficient, plastic deformation occurs at the higher temperature and the body stabilises. When it cools to ambient there is no restriction on contraction, and therefore the height is reduced whilst the width is increased in comparison with the original dimensions.

Consider welding, with the body representing the weld metal, which becomes firmly attached to the two boundaries as in Fig. 1.4.

The body is no longer free to contract from H_A to H_2 as the temperature returns to ambient. It becomes stretched and is therefore subjected to a tensile stress which will be permanent unless it is

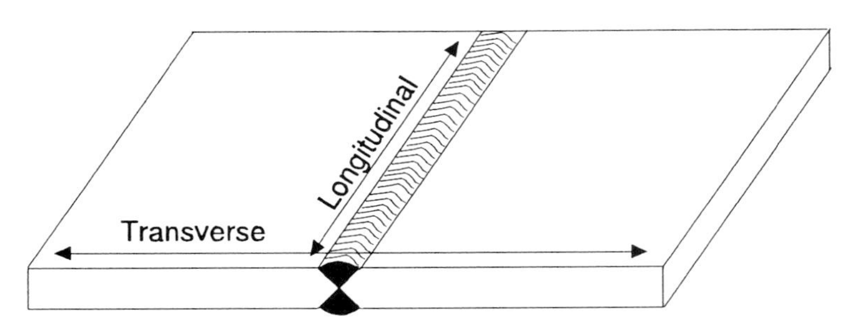

1.5 Direction of principal stresses.

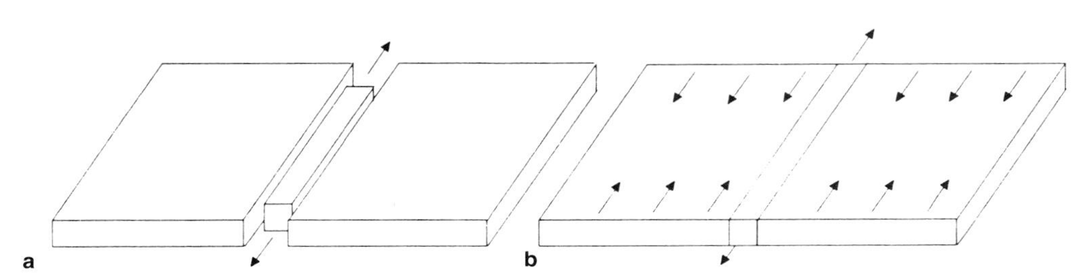

1.6 Effects of restraint on weld metal: a) Unfused; b) Fused.

relieved by some other form of treatment. Since any stress system must be in equilibrium to remain stable, compressive stresses will be developed elsewhere to give a self-balancing condition.

This description provides a very simplified explanation of the mechanism of residual welding stresses. In practice the distribution of these stresses in a welded component is complex. High tensile stresses, over yield point magnitude, will exist in the region of the joint. Consider a butt weld in an unrestrained flat plate, see Fig. 1.5. Residual stresses will act in two principal directions, longitudinal stresses parallel to the joint and transverse stresses normal to the joint.

In making the joint, gaps would occur at the plate ends if the weld metal were allowed to expand and contract without restraint, see Fig. 1.6. A longitudinal force on the weld is required to close the gap, giving a tensile stress, whilst corresponding compressive stresses in the plate material provide equilibrium.

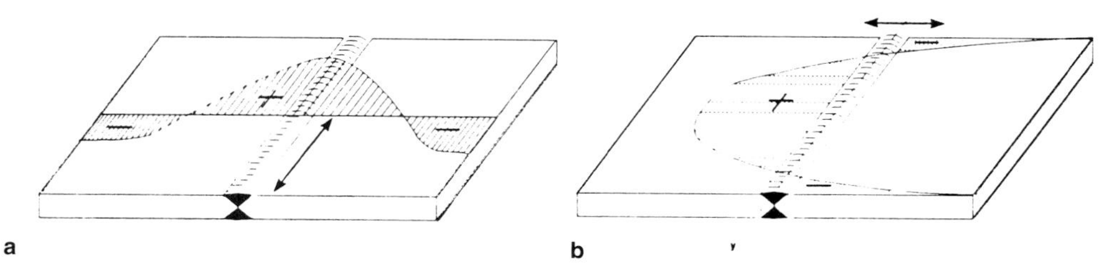

1.7 Typical residual stress distribution:
a) Longitudinal; b) Transverse.

The distribution of longitudinal residual stresses in section will be as shown in Fig. 1.7a, with the tensile component confined to the region of the joint. The corresponding transverse stress distribution is shown in Fig. 1.7b. The maximum tensile stress occurs at mid-length and to provide equilibrium, some compressive residual stresses act in the region of the joint.

It should not be forgotten that the tensile stresses can be high, often exceeding yield point magnitude.

So far the mechanical effects of welding in the form of residual stresses have been considered. Deposition of weld metal in a molten pool and localised melting of the joint faces of the components, along with subsequent cooling, all have metallurgical implications affecting the microstructure of these regions.

The optimum properties of many materials, especially steels, are derived from carefully chosen heat treatments. Steels are supplied in a variety of forms which have involved such manufacturing processes as casting, forging, rolling and extrusion. These processes involve heating to high temperature giving either a molten or plastic state which allows shaping to take place.

On completion of primary manufacturing cooling occurs, and the materials are already in some form of heat treated condition. It may be that the resultant metallurgical structure lacks homogeneity or is unsuitable for use. Further heat treatment may be necessary to develop the required properties be they tensile strength, ductility, notch toughness or creep resistance. Dependent on chemical composition, temperatures and rates of cooling are carefully controlled to create the differing microstructures associated with a required set of mechanical properties.

Cooling after welding can be relatively rapid. From the molten pool of weld metal an as-cast type of structure develops. In the region of parent metal at the fusion face heated to its melting point, metallurgical restructuring takes place. The rapid cooling results in this region having a metallurgical structure and hence mechanical properties which differ from the bulk parent material. Reheating to melting point followed by almost uncontrolled cooling has effectively undone the heat treatment processes used in manufacture. This area of parent material in the welded joint is called the heat affected zone.

Figure 1.8 shows a typical macrograph and micrographs of the metallurgical structures associated with a butt weld in carbon-manganese steel. In steels the heat affected zones are generally harder than the parent material, with corresponding loss of ductility and resistance to impact.

The likelihood of failure of a weld is a function of the severity of the thermal cycles, which lead to high residual stresses and/or weld metal and heat affected zone properties which differ radically from those of the parent material. Failure may be instantaneous, may occur after a short time has elapsed, or may manifest itself in reduced performance of the structure.

Since the basic causes of weld failure are found in thermal behaviour, a series of potential solutions arises based on application of heat. The welding process has to be controlled so that residual stresses are minimised to protect the integrity of the whole fabrication. The metallurgical structures of the weld metal and heat affected zone are controlled to give properties which are not inferior to the parent material properties used in designing the product. A series of heat treatment operations is used before, during and after welding arising from the need to control metallurgical changes and these operations form the basis of the subject of *heat treatment engineering*.

It is important to remember that prior heat treatments may have been applied to a metal during its production in order to achieve specific material properties to improve performance. A knowledge of these treatments is essential when selecting the

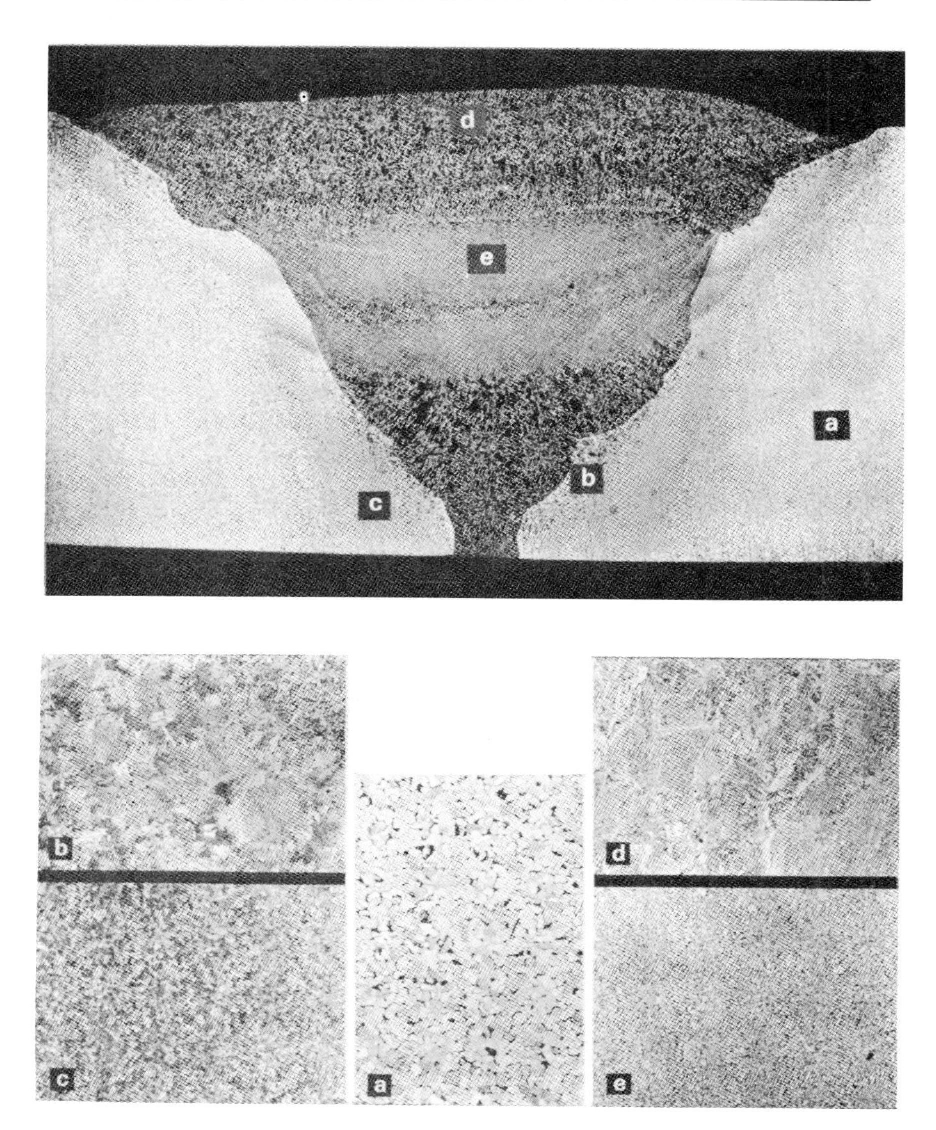

1.8 Macrosection of multi-run butt welded joint:
a) Parent plate; b) Heat affected zone at fusion face;
c) Heat affected zone in parent plate; d) Coarse grained weld metal; e) Fine grained weld metal.

operations pertinent to welding, so that material performance is not adversely affected.

Heat treatments are expensive in time and money. It is often possible to fabricate welded structures satisfactorily without heat treatment. It is also possible that heat treatment has a harmful rather than beneficial effect.

The purpose of this book is to describe the essential features of heat treatment operations, and to offer guidance as to when they are required and how to ensure that they have been correctly applied. It aims to assist designers, welding engineers and others involved in assuring the satisfactory performance of welded steel fabrications, from structural frameworks to high pressure and temperature systems.

Chapter 2

Types of heat treatment

It is often stated that the best heat treatments are those that are not necessary. Clearly the statement makes sound commercial and technical sense. Efforts are constantly being made by welding technologists and metallurgists to devise justifications for relaxing heat treatment requirements by either modifications to materials or development of welding techniques which give greater control over the factors which influence residual stress and structural change. These influences bear even greater significance in welding of complex structures, especially those which would require a localised in situ heat treatment. In such circumstances, heat treatment after welding to reduce residual stresses poses problems of practicality and desirability. The thermal gradients inherent in such an approach can induce further stresses large enough to affect the integrity of the structure.

However, except in exceptional circumstances, heat treatment before, during and after welding remains an essential part of fabrication, dependent on material and service. The treatments associated with welding of steel structures are described below.

Outgassing or hydrogen bake-out

Hydrogen can be regarded as an enemy in welding ferritic steels because of its potential embrittling effect, which can cause cracking. If it is necessary to weld steels which have been exposed to hydrogen or hydrogen bearing substances, then precautions are necessary before welding.

Outgassing or bake-out can be a preparatory operation applied to operational plant exposed to hydrogen bearing products during service, which is to undergo weld repair or modification. The objective of this treatment is to raise the temperature of the fabrication and hold, to allow hydrogen to diffuse out before welding, and so reduce susceptibility to this mode of cracking. Temperature and hold times vary dependent on steel classification and section thickness, with values up to 450 °C and between 1 and 48 hours respectively being known, the higher temperatures being held for the shorter soak times. These temperatures and times will not be sufficient to change the microstructure or residual stress state.

The usual methods are based on electrical resistance heating, as used for localised heat treatments, and thermocouples are used to monitor and control the process. A time/temperature chart is produced to provide the necessary assurances and record that the cycle has been carried out.

Preheat

Preheating involves raising the temperature of the parent material locally on both sides of the joint to a value above ambient. The need for preheat is usually determined by the pertinent fabrication code and verified by the weld procedure qualification test. Preheat may be required as an aid to welding for one of four basic reasons:

(a) *To slow the rate of cooling, especially below about 800 °C in the heat affected zone, to reduce hardness.* High carbon and low alloy steels harden if they are quenched from high temperatures (above about 850 °C, hotter than cherry red). Exactly the same process can happen in a welded joint at the fusion face with the parent material. By raising the temperature of the base metal to be welded, to reduce the temperature differential between the weld pool and the surrounding base material, hardening may be reduced as the weld cools. Reducing hardness reduces the risk of cracking.

(b) *To control the diffusion rate of hydrogen in a welded joint.*

The welding arc breaks down water, present as moisture, into its basic elements of hydrogen and oxygen. Both of these gases are easily absorbed into the weld metal at high temperatures and can remain trapped during cooling. At lower temperatures hydrogen can play an important role in weld and heat affected zone cracking – known as hydrogen or cold cracking. Preheat, with its associated beneficial effect on cooling rate, helps to promote diffusion of hydrogen out of the steel structure. Preheat can also help by ensuring that the weld preparation area is dry and remains dry throughout welding.

Moisture is also introduced by the welding consumables, being present in electrode coatings and fluxes. To obtain the maximum benefits from preheat in controlling hydrogen, it must be accompanied by careful controls over removal of moisture from the welding consumables by following manufacturers' baking and storage instructions.

(c) *To reduce thermal stresses.* Thermal strains are set up as the molten weld pool cools, as described in Chapter 1. Partially made welds can crack as the parent metal opposes or restrains the contraction of the weld metal and the cross sectional area of the joint is insufficient to withstand the resultant tensile stress. Preheat can control the level of strain by reducing temperature differentials and reducing cooling rates.

(d) *Compensation for heat losses.* Thicker section steels with high thermal conductivity may need preheat during welding to ensure fusion. Preheat may be required for thinner sections for the other reasons discussed.

Preheat is expensive and should be applied only when it is essential to complete a joint satisfactorily. Use of a higher preheat temperature than necessary is a costly and over-conservative practice with dubious advantages and should be avoided. However, when preheat is required, every effort should be made to ensure that the correct temperature for the application is reached, so that preheat is effective over the region being welded and maintained for the duration of the welding process.

Type of steel	Minimum preheating temperature for TIG welding of root runs, °C			Hydrogen controlled weld metal (note 1)		Non-hydrogen controlled weld metal	
	Carbon steel root run		Matching root run, all diameters and thicknesses	Material thickness, mm	Minimum preheating temperature, °C	Material thickness, mm	Minimum preheating temperature, °C
	≤127 mm dia. and ≤12.5 mm thick	>127 mm dia. or >12.5 mm thick					
C and C-Mn, (≤0.25 C)	5	≤30 mm*:5 >30 mm*:100	≤30 mm*:5 >30 mm*:100	≤30 >30	5 100	≤20 >20	5 100
C and C-Mn, (> 0.25 C ≤ 0.4 C)	50	100	100	All	150	All	200
C-Mo	5	100	≤12.5 mm*:20 >12.5 mm*:100	≤12.5 >12.5	20 100	≤38 (note 5)	150
1 Cr ½ Mo 1¼ Cr ½ Mo	5	100	100	≤12.5 >12.5	100 150 } (note 3)	≤12.5 > 12.5 ≤ 20 (note 5)	150 200
½ Cr ½ Mo ¼ V	50	100	100	≤12.5 >12.5	150 200 } (note 3)	Note 6	Note 6
2¼ Cr 1 Mo	50	100	100	≤12.5 >12.5	150 200 } (note 3)	≤12.5 (note 5)	200
5 Cr ½ Mo 7 Cr ½ Mo 9 Cr 1 Mo	Note 2	Note 2	150	All	200 (note 3)	Note 6	Note 6
12 Cr Mo V(W)	Note 2	Note 2	150	All	150 (note 4)	Note 6	Note 6
3½ Ni 9 Ni	Note 2	Note 2	100	All	150	Note 6	Note 6

NOTE 1: Hydrogen controlled weld metal as defined in BS639, contains not more than 15 mL of diffusible hydrogen per 100 g of deposited metal when determined by the method given in BS6693: Part 2.
NOTE 2: A carbon steel root run is not to be used.
NOTE 3: When TIG welding is used a lower pre-heating temperature may be applied provided it is proved to be satisfactory by procedure tests.
NOTE 4: A welding process with extra low hydrogen potential (less than 5 mL per 100 g of deposited weld metal) should be employed. If a high preheating temperature is used, e.g. 300 °C, then the joint should be cooled to between 100 °C and 150 °C to ensure full transformation before post-weld heat treatment is applied. Welding at a preheating temperature in the region of 150 °C to 200 °C will ensure that a high degree of transformation and some tempering will occur during welding and will assist in the removal of hydrogen from the joint.
NOTE 5: Above the maximum thickness stated hydrogen controlled weld metal only to be used.
NOTE 6: Hydrogen controlled weld metal only to be used.
* Thickness.

2.1 Guidance on preheat in national codes: a) From BS2633:1987.

Guidance on the need to preheat is generally obtained from national fabrication codes, which list recommended minimum temperatures for steel types grouped by composition and also give the minimum section thickness to which they apply. The

Base metal P-No. [Note (1)]	Weld metal analysis A-No. [Note (2)]	Base metal group	Nominal wall thickness		Specified min. tensile strength, base metal		Min. temperature			
							Required		Recommended	
			in	mm	ksi	MPa	°F	°C	°F	°C
1	1	Carbon steel	<1	<25.4	≤71	≤490	...	...	50	10
			≥1	≥25.4	All	All	...	...	175	79
			All	All	>71	>490	...	...	175	79
3	2, 11	Alloy steels, Cr ≤ ½%	<½	<12.7	≤71	≤490	...	...	50	10
			≥½	≥12.7	All	All	...	...	175	79
			All	All	>71	>490	...	...	175	79
4	3	Alloy steels ½% < Cr ≤ 2%	All	All	All	All	300	149	...	...
5	4, 5	Alloy steels, 2¼% ≤ Cr ≤ 10%	All	All	All	All	350	177	...	...
6	6	High alloy steels martensitic	All	All	All	All	...	...	300[3]	149[3]
7	7	High alloy steels ferritic	All	All	All	All	...	...	50	10
8	8, 9	High alloy steels austenitic	All	All	All	All	...	...	50	10
9A, 9B	10	Nickel alloy steels	All	All	All	All	...	...	200	93
10	...	Cr-Cu steel	All	All	All	All	300–400	149–204	...	...
10A	...	Mn-V steel	All	All	All	All	...	...	175	79
10E	...	27Cr steel	All	All	All	All	300[4]	149[4]	...	...
11A SG 1	...	8Ni, 9Ni steel	All	All	All	All	...	...	50	10
11A SG 2	...	5Ni steel	All	All	All	All	50	10	...	...
21–52	...	...	All	All	All	All	...	...	50	10

NOTES:

(1) P-Number from BPV Code, Section IX, Table QW-422. Special P-Numbers (SP-1, SP-2, SP-3, SP-4, and SP-5) require special consideration. The required thermal treatment for Special P-Numbers shall be established by the engineering design and demonstrated by the welding procedure qualification.

(2) A-Number from BPV Code, Section IX, QW-442.

(3) Maximum interpass temperature 600 °F (316 °C).

(4) Maintain interpass temperature between 350 °F–450 °F (177 °C–232 °C).

Fig. 2.1 b) From ASME B31.3:1990.

significance of the codes and standards which govern welding is discussed in Chapter 7. However, for the purpose of illustration, Fig. 2.1 shows the preheat requirements of two high pressure pipework codes, BS2633 and ANSI B31.3 respectively.

The preheat temperatures and thickness criteria vary between these two examples for a given material type, but it is clear that as the alloying content of a ferritic steel increases, so does the significance of preheat. This is associated with an increased risk of cracking due to hardenability without control over cooling.

For general welding of carbon-manganese steels, the appropriate guidance document for setting up welding procedure trials is BS5135 (pr EN1011). This takes into account the chemical analysis of the steel in the form of carbon equivalent, C_{eq}, expressed as:

$$C_{eq} = C + \frac{Mn}{6} + \frac{Cr + Mo + V}{5} + \frac{Cu + Ni}{15}$$

the combined thickness of the joint (heat sink effect), the energy input from the welding process (kJ/mm = [arc volts × amps] / travel speed), and the hydrogen potential of the electrode coatings (scale A > 15 ml/100 g to scale D < 5 ml/100 g). An example of a chart to determine preheat from the combination of these variables is shown in Fig. 2.2. (See Chapter 3 for more details.)

Preheat temperatures can range from nominal heating only, where structures reach sub-zero temperatures, or may be as high as 250 °C. It is normal practice to maintain preheat temperature at a minimum level throughout welding, although heat input may increase the temperature to such a level that mechanical properties may be impaired or welding conditions altered so as to introduce defects. Accordingly, a maximum interpass temperature may also be specified, usually some 100–150 °C above the minimum preheat level.

Post-heat

This is the term given to an extension of preheat on completion of welding to maintain the same or an increased temperature. Its purpose is to effect diffusion of hydrogen from the joint and reduce susceptibility to the associated form of cracking. It is usually applied to higher strength carbon-manganese steels and low alloy steels where the risk of hydrogen cracking is higher.

Guidance on post-heat treatments is not commonly given in

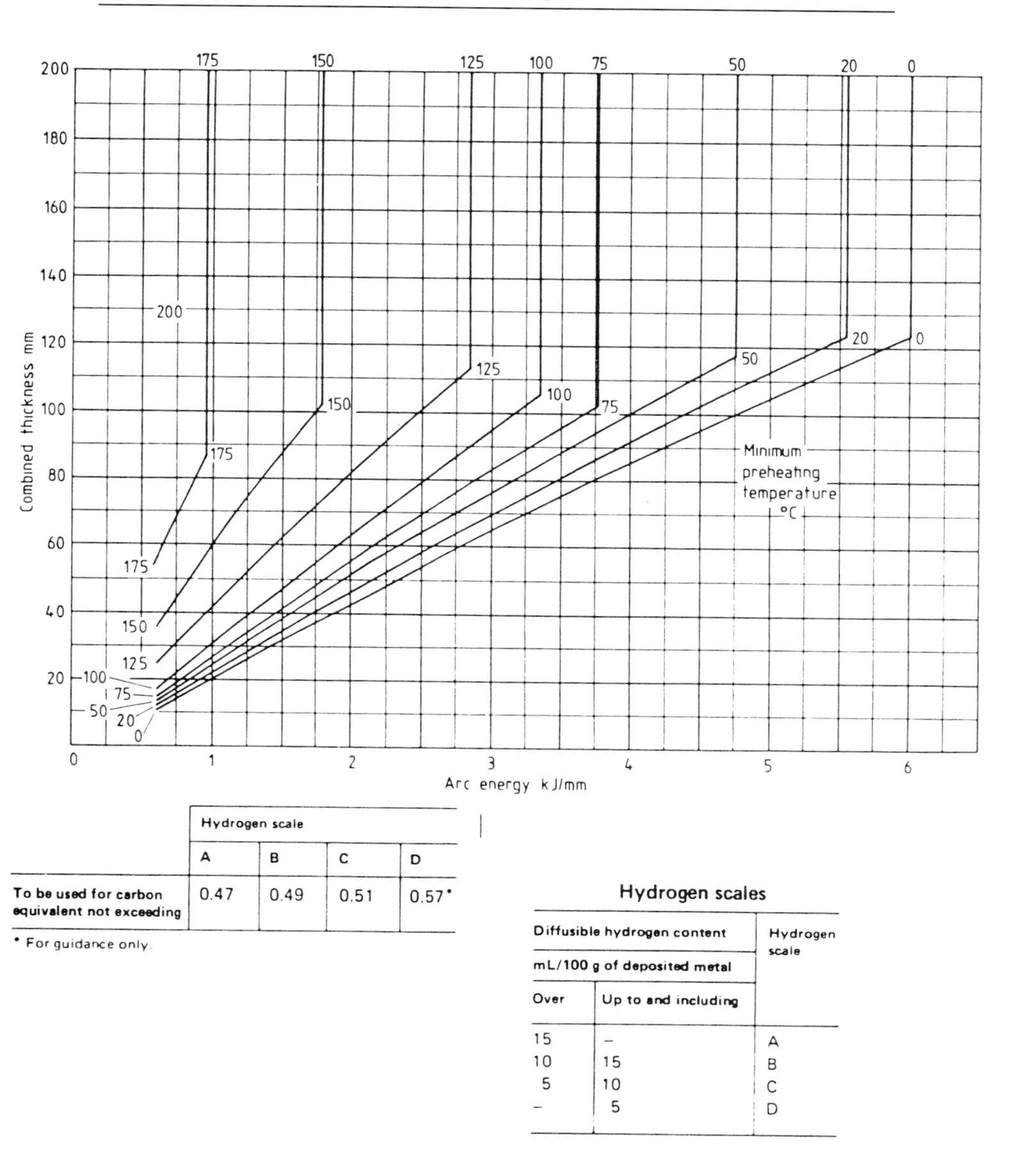

	Hydrogen scale			
	A	B	C	D
To be used for carbon equivalent not exceeding	0.47	0.49	0.51	0.57*

* For guidance only.

Hydrogen scales

Diffusible hydrogen content mL/100 g of deposited metal		Hydrogen scale
Over	Up to and including	
15	–	A
10	15	B
5	10	C
–	5	D

2.2 A typical nomogram from BS5135 (pr EN1011).

national standards or codes, but post-heat is often specified by clients who have incorporated their requirements into the weld procedure qualification test. Temperatures and soak times are derived from numerous technical papers, with guidelines from TWI being a popular source of information. A requirement for post-heat immediately implies the need for a fully controlled preheat.

Post-weld heat treatment

This is the process commonly referred to as stress relief, so called because it is carried out at temperatures at which the yield strength has fallen to a low value. If the structure is heated uniformly, the yield strength of the material around the weld is unable to support the initial level of the residual stresses, which are relieved by plastic deformation. Creep occurs at elevated temperatures and strain occurs by a diffusion mechanism, relaxing the residual stresses even further. The extent to which residual stresses are relaxed depends on temperature and time for any given material and on material for any given temperature, see Fig. 2.3. A common guideline on time for post-weld heat treatment is that the joint should be soaked at peak temperature for a period of 1 hour for each 25 mm (1 inch) of thickness.

The stress distribution at the higher temperatures becomes more uniform and stress reduces to a low level. After cooling, provided that it is carried out in a controlled manner, the improved stress distribution is retained.

In addition to a reduction and re-distribution of residual stresses, post-weld heat treatments at higher temperatures permit some tempering or ageing effects to take place. These metallurgical changes are usually very beneficial in that they reduce the high hardness of the as-welded structures, improving ductility and reducing the risks of brittle fracture. In some steels, ageing/precipitation may cause toughness and ductility to deteriorate, in which case specialist advice should be taken on times and temperatures for different steels.

To be able to control the metallurgical changes beneficially, an upper limit to the soak temperature is necessary to avoid

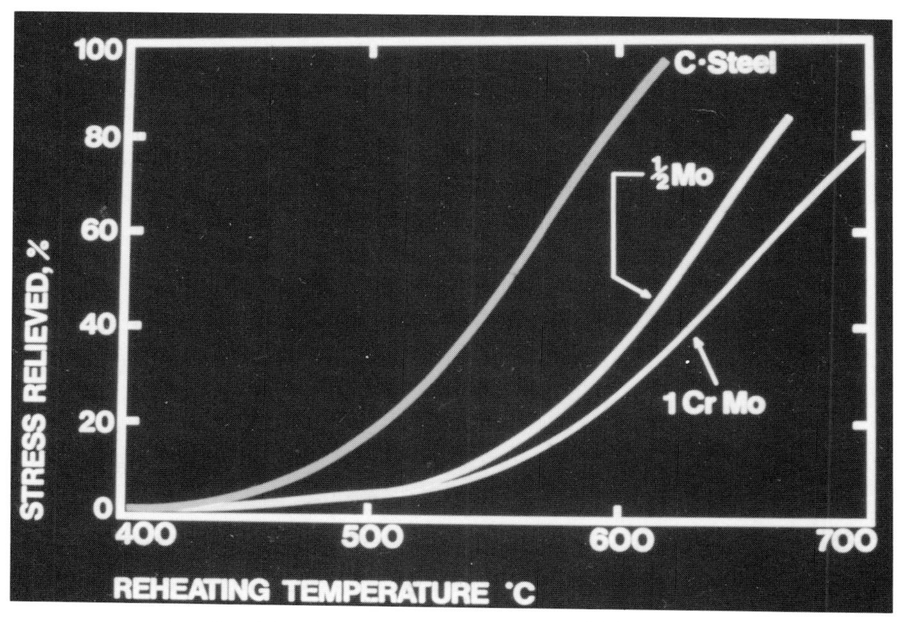

2.3 Residual stress versus post-weld heat treatment temperature curve.

impairing the mechanical properties of the parent material. This requires a knowledge of the equilibrium diagram, notably the lower critical (Ac_1) line (Fig. 2.4).

Post-weld heat treatment is mandatory in some national standards and codes, as well as being required to offer acceptable component life in onerous environments. As with preheat, the alloying content of the steel is related to post-weld heat treatment temperatures, as discussed in Chapter 7.

The necessity for post-weld heat treatment depends on the material and service requirements. Other factors include welding conditions and a knowledge of types of failure. As post-weld heat treatments involve much higher temperatures, costs are higher than for preheat and post-heat. Again, consideration should be given to use of post-weld heat treatment only if it is necessary to ensure that the service performance of the structure meets the design criteria.

It is a more critical process with increased potential for

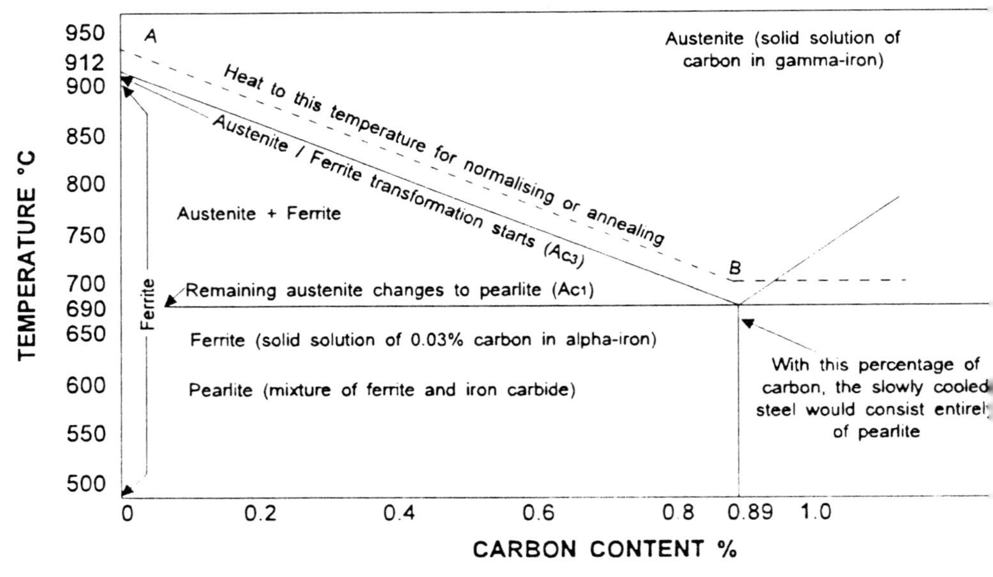

2.4 Equilibrium phase diagram for mild steel.

detrimental effects if not carefully carried out. This necessitates careful assessment of heating rates, cooling rates, tolerances on holding temperature and holding time if benefits are to be ensured.

Features of post-weld heat treatment

Five aspects of post-weld heat treatment must be addressed:

1. The hot zone should be adequate to raise the weldment to the required temperature and provide a uniform temperature profile to avoid creating additional undue thermally induced stresses. This aspect has greater significance in localised heat treatments, but nevertheless must also be considered in furnace heat treatments.
2. The heating and cooling rates should be at least compliant with the necessary code requirements. These rates will indicate absolute maximum values. The rates are calculated from

simple formulae related to component thickness and offer protection against thermally induced stresses. With thicker and more complex structures an experienced heat treatment engineer may wish to consider lower rates than required by the code to ensure acceptable temperatures profiles and gradients, with a view to keeping thermally induced stresses to an absolute minimum.

3. In localised heat treatment, the temperature gradients away from the hot zone should not be unduly severe, again the objective being minimisation of thermally induced stresses. British Standards BS5500 and BS2633 offer guidance on this issue, quoting the $2.5\sqrt{Rt}$ rule.
4. The soak temperatures should be held within the upper and lower limits of the soak range for the appropriate time.
5. The heat treatment system (including insulation), zonal division and number of thermocouples should be such that the energy input and level of control are capable of enabling these objectives to be met, ensuring that the integrity of the overall structure is not jeopardised.

Where furnace heat treatments are involved, then provided that the unit is efficient, control of the heat treatment depends on establishment of the requisite heating cycle determined with due respect to the component geometry. Reliance is placed on the accuracy with which the energy medium is controlled and the temperatures are recorded.

In local heat treatments, controls have to be implemented to give assurance that the engineered system can provide a suitable performance.

Normalising

This heat treatment involves higher temperatures than post-weld heat treatment, typically 850–950 °C for ferritic steels, which is above the upper critical (Ac_3) line on an equilibrium diagram (Fig. 2.4). Cooling takes place in still air which results in a refinement of the metallurgical grain size compared with the condition before normalising. The finer grain size leads to increased yield strength and better resistance to fracture. For certain types of steel more precise control over mechanical properties can be obtained by subsequently tempering at temperatures below the lower critical (Ac_1) line.

Chapter 3

Effects of heat treatment

Preheat

Considering most steels used in welded structures as belonging to one of three main groups, the following notes summarise the importance of preheat and its benefits.

Carbon-manganese steels

Under certain conditions, C-Mn steels are susceptible to a form of cracking[1] in the HAZ of a weld which is known by various names, such as hydrogen induced cracking, cold cracking, underbead cracking, or HAZ cracking. Hydrogen may be present in deposited weld metal as a result of breakdown of moisture in consumables, such as in electrode coatings. The term cold cracking is used to indicate that this form of cracking occurs only when the joint has nearly or completely cooled to room temperature; not infrequently cold cracking may occur up to two days after the joint has been made, and occasionally longer.

In many cases cold cracking can be prevented by application of preheat, which is beneficial for two reasons. First, the lower cooling rate after welding may prevent formation of susceptible microstructures in the HAZ. Secondly, the preheat maintains the HAZ at a temperature where embrittlement is reduced by giving time for the hydrogen level to drop by diffusion away from the joint. This form of cracking occurs only in susceptible microstructures in the presence of hydrogen and under tensile stress. For C-

Mn steels a reasonable indication of the presence of such microstructures is given by hardnesses exceeding 350 HV, which corresponds to formation of microstructures containing martensite and sometimes bainite.

Transformation to microstructures having hardnesses greater than 350 HV can begin at temperatures of 550 °C and continue to 200 °C, depending on composition, and a good correlation has been found between susceptibility to cracking and cooling rate from 800 to 500 °C. Thus, preheat temperatures up to 300 °C will have a marked effect on the cooling rate in the critical range and therefore on the possibility of cracking. The macrograph in Fig. 3.1 shows a typical hydrogen induced crack within the heat affected zone of a C-Mn steel.

In addition, by maintaining joints at temperatures above ambient, susceptibility to cracking is reduced and sufficient time may

3.1 Hydrogen cracking in the heat affected zone of a carbon-manganese steel.

be permitted for hydrogen in the joint to diffuse out.[2] The response to varying cooling rates of steels in the C-Mn range can be conveniently expressed in terms of carbon equivalent, C_{eq}, formulae, in which the chemical composition of the steel is expressed as the sum of effects of alloying elements on hardenability. Several formulae have been published, but the most commonly used version for C-Mn steels, including allowance for residuals, is (see Chapter 2 also):

$$C_{eq} = C + \frac{Mn}{6} + \frac{Cr + Mo + V}{5} + \frac{Cu + Ni}{15}$$

The formula must be used with caution, and particular attention is drawn to the requirements of BS5135 (pr EN1011) – British Standard specification for *Arc Welding of Carbon and Carbon Manganese Steels*. Through a series of tables and nomograms this document correlates carbon equivalent, arc energy, combined thickness, hydrogen content and preheat level, and gives combinations of these factors which will avoid hardened microstructures in single run welds for these steels. A typical nomogram was shown in Fig. 2.2. However, the factors associated with avoidance of hydrogen cracking are constantly being investigated, leading to recommendations on how the accuracy of the standard can be improved.[3]

For a given carbon equivalent and diffusible hydrogen content (see Fig. 2.2), values of either combined thickness, arc energy or minimum preheat temperature can be determined given any two of these three factors. The combined thickness of a joint is a measure of the paths available for heat flow away from it, see Fig. 3.2. Thus for a butt weld in 40 mm thick plate, the combined thickness is 80 mm. For a fillet weld between a 13 mm thick web and a 40 mm thick base plate the combined thickness is 93 mm. Arc energies are obtained from a series of tables contained in the standard, expressed in terms of electrode size, weld run length and run out rates.

Further information on the correct approach to welding can be obtained from the second edition of *Welding Steels without Hydrogen Cracking*.[4] This book is more detailed than BS5135 and recognises factors not covered by the standard, principally

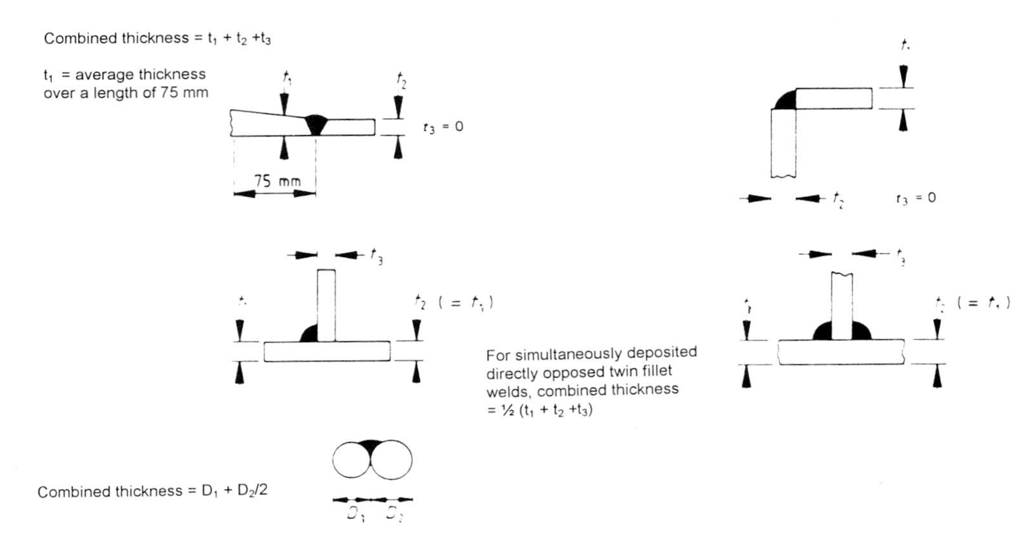

3.2 Determination of combined thickness (after BS5135:1984).

mechanical influences such as restraint, joint type and fit-up. The methodology is to compensate by assuming a more stringent condition for diffusible hydrogen. For certain modern steel grades, even further considerations are necessary to avoid hydrogen cracking.[5]

Both BS5135 (pr EN1011) and the original *Welding Steels without Hydrogen Cracking* by F R Coe, have been combined by TWI to produce a software package PREHEAT – a microcomputer programme to help avoid heat affected zone (HAZ) hydrogen cracking when welding carbon and carbon-manganese steels. There are four principal inputs: carbon equivalents, arc energy, preheat level and hydrogen content. If any are unknown, the responses to a series of questions may enable values to be established from other information. The function of the software is to determine feasible combinations of these factors.[6]

For example, carbon equivalents can be obtained given the steel grades to BS EN 10025 or BS EN 10113, arc energies can be deduced knowing the welding parameters. The programme is versatile in that alternative nomograms can be produced dependent on the known and unknown principal inputs, see Fig. 3.3. The

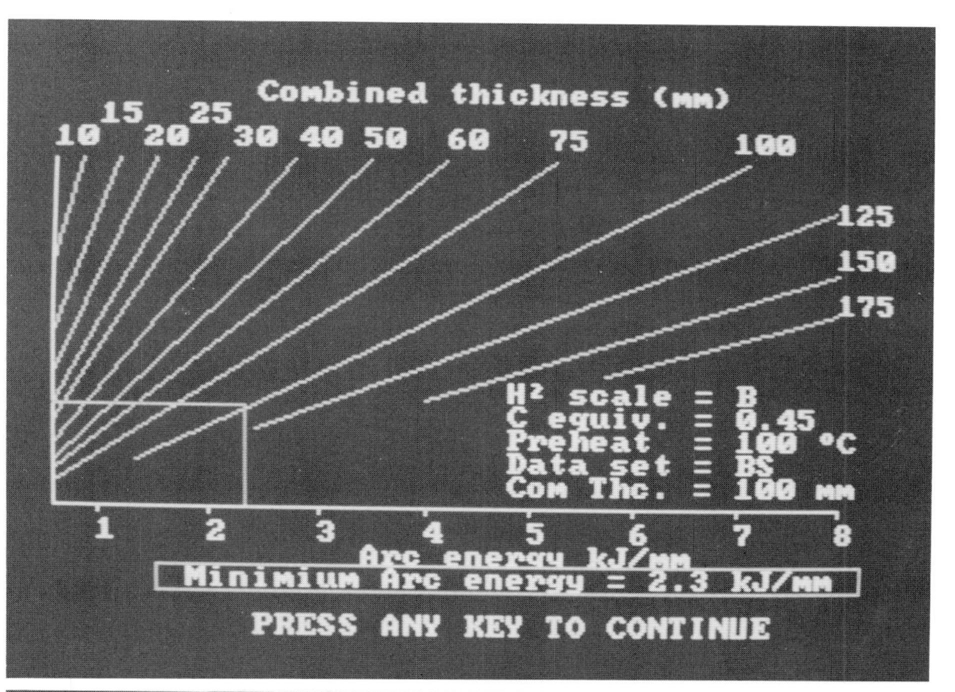

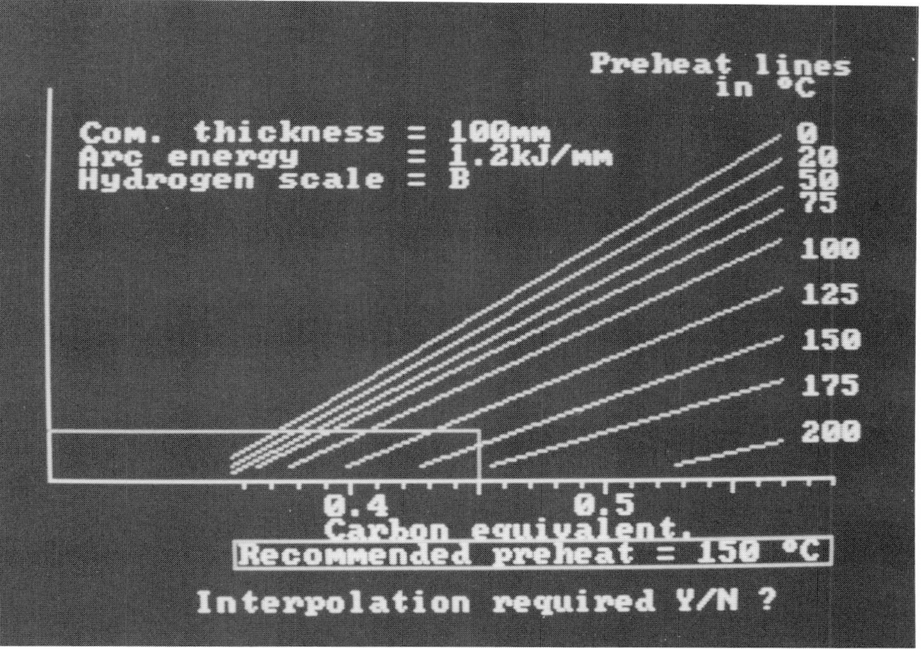

3.3 Nomograms generated by TWI's PREHEAT software (courtesy TWI).

effects of varying parameters can be examined before deciding on a weld procedure.

A print out from the software can provide quality assurance documentation, see Fig. 3.4. The software assists those who are not familiar with the British Standard, reducing the likelihood of selecting the wrong nomogram.

Whilst the above sources of information relate to single-run welds, in multi-run welded joints there is usually an appreciable tempering effect of later runs on those first deposited. This means that the full hardness of a single-run weld is reached only if the last run is made adjacent to plate material. In addition, there is now appreciable evidence that it is possible to tolerate microstructures with a hardness greater than 350 HV, provided that the hydrogen content is very low. Thus if the preheat is high enough and is maintained for long enough it may permit diffusion of initial higher hydrogen concentrations away from the joint, producing a low enough level to cool safely without cracking.

In some cases use of high preheat has to be considered in conjunction with requirements for maximum interpass temperature. An increase in interpass temperature may be accompanied by a decrease in weld metal strength because softer microstructures are formed by the lower cooling rate. This may impose a limitation on interpass temperature, for example when maximum fracture toughness is required in the HAZ. Adherence to an interpass temperature requirement can increase welding time, but for critical applications its significance cannot be ignored.

Thus, it is recommended that procedural tests be carried out to establish the preheat conditions necessary to avoid cracking whilst achieving optimum weld performance, and that the previously referenced sources should be used as a basis for deciding the parameters for procedural tests.

Low-alloy steels

The problem of HAZ cracking described for C-Mn steels may be more acute in low-alloy steels. These are usually more hardenable

```
**********************************************************************
*                                                                    *
* TWI                                                           1985 *
*                                                                    *
**********************************************************************
*                                                                    *
* PREHEAT                 for carbon and carbon-manganese steels     *
*                                                                    *
**********************************************************************

   PROCEDURE REFERENCE :  WPP 324
   DATE                :  15/3/1994

---------------------------------------------------------------------

   DATA SET               :   British Standard

   PROCESS                :   MIG

   WELD TYPE              :   Butt

   COMBINED THICKNESS     =   80 mm

   CARBON EQUIVALENT      =   0.41

   HYDROGEN SCALE         :   B

   ARC ENERGY             =   1.0 kJ/mm

----------------------------------------------------------------------
 RECOMMENDED PREHEAT    = 125 deg C
----------------------------------------------------------------------

---------------------------------------------------------------------
   PREHEATING INSTRUCTIONS   :-

   1     This is the temperature of the parent material immediately
         prior to welding.

   2     The parent material temperature must reach 125 deg C at
         least 75mm away from the weld location on each joint member.

   3     The temperature  should be measured on the opposite face
         to that being heated.

   4     If there is access to one face only,then after removing
         the heat source,time should be allowed for temperature
         equalisation before measuring (2 minute per 25 mm of plate
         thickness).

   5     The minimum interpass temperature will also be 125 deg C,
         except where the filling passes utilise a higher arc energy
         than the root run.

   6     The heat is assumed to be locally applied . General
         preheating of the specimen may allow a reduction in the
         required preheat.
---------------------------------------------------------------------
```

3.4 A procedure produced by PREHEAT software (courtesy TWI).

than C-Mn steels because of the presence of small amounts of additional alloying elements, and usually have increased yield and ultimate strength compared with C-Mn steels. Low-alloy steels can be supplied in a heat-treated condition so as to develop optimum properties, usually for improved service at elevated temperatures.

In general, higher preheats are necessary to prevent cracking for a given thickness and given welding conditions than for C-Mn steels. Attempts have been made to derive carbon equivalent formulae which can be used in the same way as for C-Mn steels, but this approach is not entirely successful since the interaction between different elements does not lend itself to this form of expression. The formula given for C-Mn steels can be used as a rough guide but inconsistencies will occur. It is usually necessary to confirm the required preheat, either from general experience, or by carrying out a welding procedure test in which a test plate is sectioned and examined for cracks, and the hardness determined.

The critical hardness of 350 HV often quoted for C-Mn steels, can be applied to other ferritic steels as a reasonable guide to susceptibility to cracking, but it should not be regarded as an absolute criterion.[7] The critical hardness is dependent upon the hydrogen level of the welding process, process energy, selected preheat (and interpass), restraint and operating stresses, fit-up and actual composition of the steel being welded.[4]

In steels with an increased alloy content it may not be possible to use preheat to avoid forming hardened microstructures in the HAZ.

Whilst control of hydrogen content, such as by use of low-hydrogen consumables, is necessary to prevent cracking, there are sometimes circumstances in which spontaneous cracking would occur if a joint were permitted to cool to room temperature in the as-welded condition. In such materials, cracking can be prevented by preheating to temperatures of 200–300 °C, followed by an immediate post-weld heat treatment to 600–750 °C, to soften the HAZ before it is allowed to cool to room temperature. Cr–Mo–V steels fall into this category as do 12%Cr steels which have particularly hard and brittle as-welded structures. Where experience is not available to give guidance on steels of particular compositions, procedural tests will usually be necessary.

Austenitic steels

Hydrogen cracking does not occur in austenitic steels and preheat is not required nor used to avoid this problem, although it may have to be considered for mixed ferritic/austenitic weldments if required by the ferritic material. Austenitic steel weldments, however, may be subject to a form of HAZ cracking during elevated temperature service or stress relief. Preheating to high temperatures could reduce the cooling rate following welding to such an extent that precipitation occurs during cooling in a form which prevents subsequent embrittlement during reheating.

Preheats at this level are very difficult to apply in practice owing to the intense discomfort caused to the welder, although preheats of up to 500 °C have been used to weld certain types of highly alloyed material (17%Cr).

Duplex austenitic–ferritic steels

Preheat, interpass temperature and thickness of section to be welded influence the austenite–ferrite balance in duplex steels. Low preheat and controlled interpass temperatures can contribute towards cooling rates which provide acceptable microstructures. It is suggested that preheat temperatures for duplex steels are restricted to between 10 and 20 °C with an interpass limitation of 150 °C. For welding super duplex stainless steels, lower interpass temperatures are recommended.[8]

The requirements of a number of standards for design and fabrication of pressure vessels and pipe systems, which identify requirements for preheating of steels to be welded, are given in Chapter 7.

Post-heat and post-weld heat treatment

The significance of post-weld heat treatment or stress relief for a broad categorisation of steel types is summarised below.

Carbon-manganese steels

Maintenance of post-heat after welding as a continuation of pre-heat and interpass temperatures is not usually required for C-Mn steels unless the carbon and manganese reach high levels so that there are increased risks of hydrogen cracking. Again this question should be settled by procedural tests, but consideration should be given to use of post-heat for carbon equivalents exceeding 0.6%. The most commonly applied post-weld heat treatment for these steels is stress relief.

Thermal stress-relief treatments can be beneficial in improving resistance to stress corrosion and brittle fracture as discussed later. In some codes of practice, stress relief of welded constructions is mandatory, for example Class 1 pressure vessels to BS5500. There are many constructions, however, in which stress relief heat treatments are not practicable because of the size and nature of the fabrication. The need for stress relief becomes greater with increased thickness of material and increased size of welded joints. In highly restrained joints in thick material cracking may develop in the welded joint under added stress from service loading, and this cracking can best be prevented by a final stress-relief treatment after welding. Intermediate stress-relief treatments may occasionally be useful, but only if no other means can be established to prevent cracking.

For general use, C-Mn steels are not particularly susceptible to stress corrosion effects, although if the welding conditions produce a martensitic heat affected zone it is possible for a stress corrosion form of hydrogen cracking to develop when hydrogen is present, as in sour gas pipelines. For other reasons C-Mn steels are not commonly used for high pressure hydrogen containing pressure vessels, and the problem of hydrogen cracking in service does not commonly arise. When caustic soda, nitrate solutions, or sodium chloride are present in contact with C-Mn steel weldments the possibility of stress corrosion cracking arises. Stress relief heat treatments are beneficial in such cases in that they reduce the total level of stress close to welded joints and reduce the likelihood of stress corrosion cracking.[9]

The effect on fatigue behaviour of relieving residual stresses is

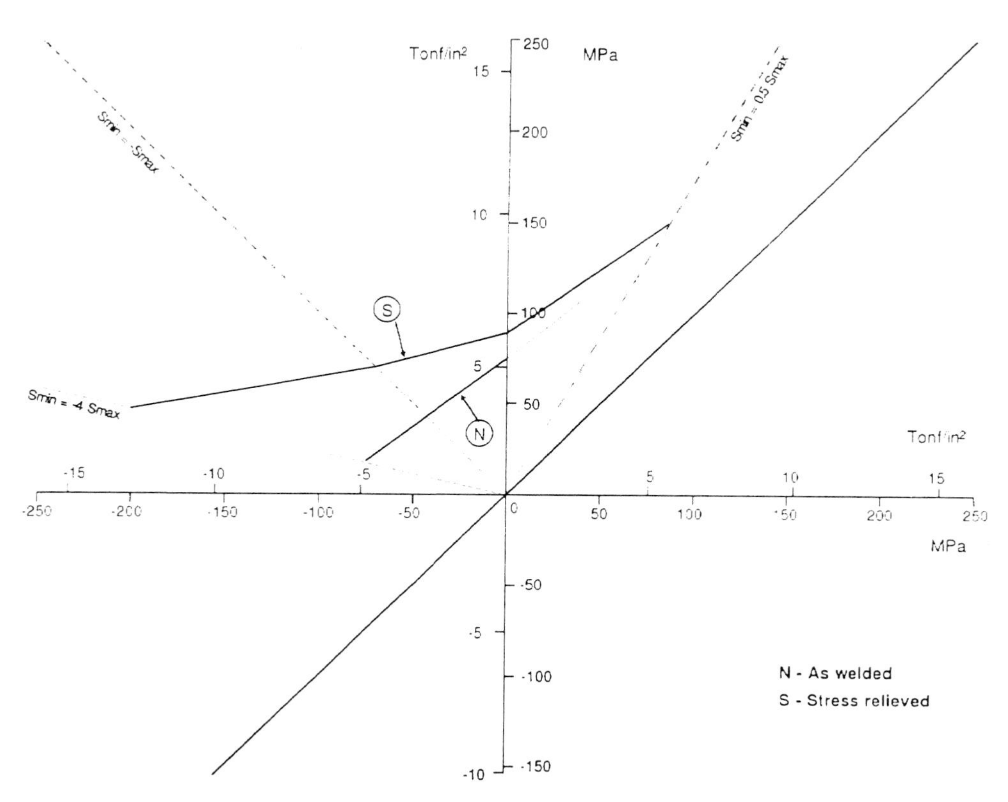

3.5 Modified Goodman diagram showing effect of stress relief on fatigue behaviour.

secondary compared with the inherent fatigue strength of particular welded details. Little or no benefit is achieved in terms of resistance to fatigue by stress relief heat treatment in those cases in which a large part of the stress range is tensile. Some benefit is obtained when the major part of the stress cycle is compressive. This can be seen from modified Goodman diagrams for different welded details, see Fig. 3.5.[10]

Perhaps the most striking benefit of stress relief heat treatment arises where there is a risk of brittle fracture. In as-welded fabrications there may be regions of the heat affected zone of a weld which have much impaired fracture toughness compared with the parent plate. This may arise, for example, by mecha-

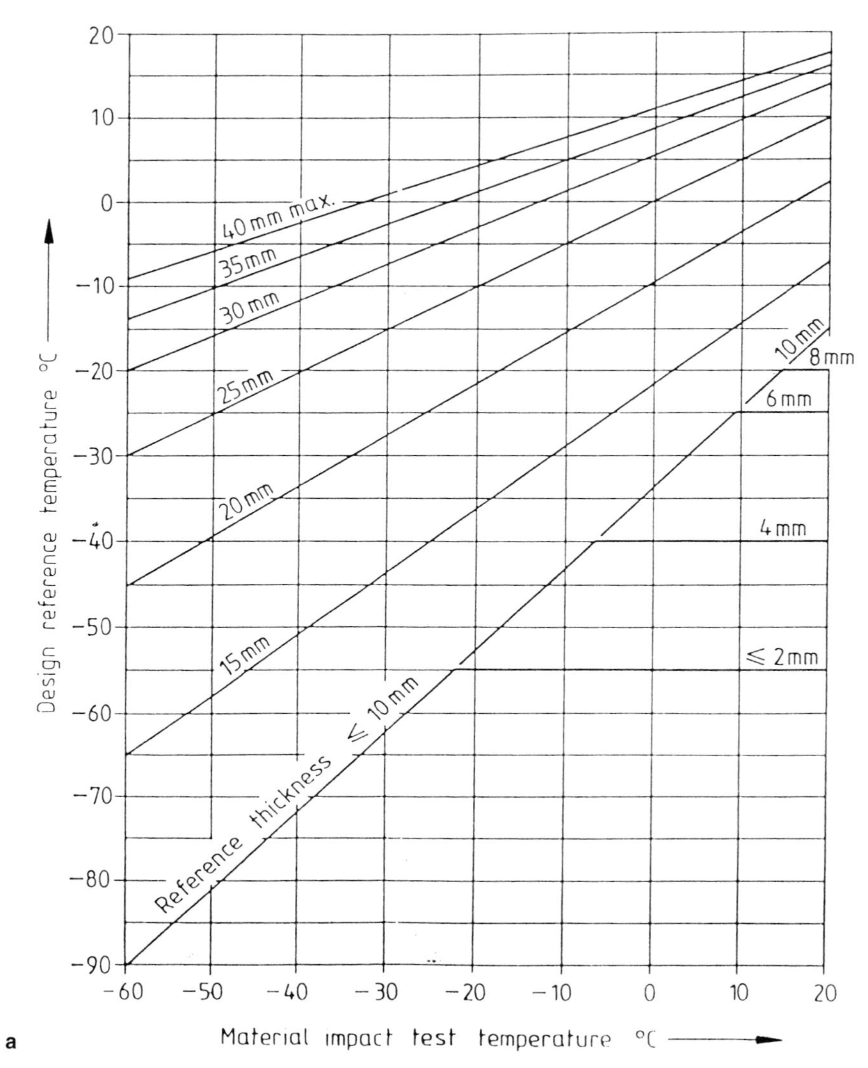

3.6 Design temperature/thickness/impact test temperature diagrams: a) As-welded components.

nisms of transformation in the heat affected zone or by strain ageing.

In addition to the obvious effects on residual stresses, heat treatment can produce tempering of hardened zones and over-

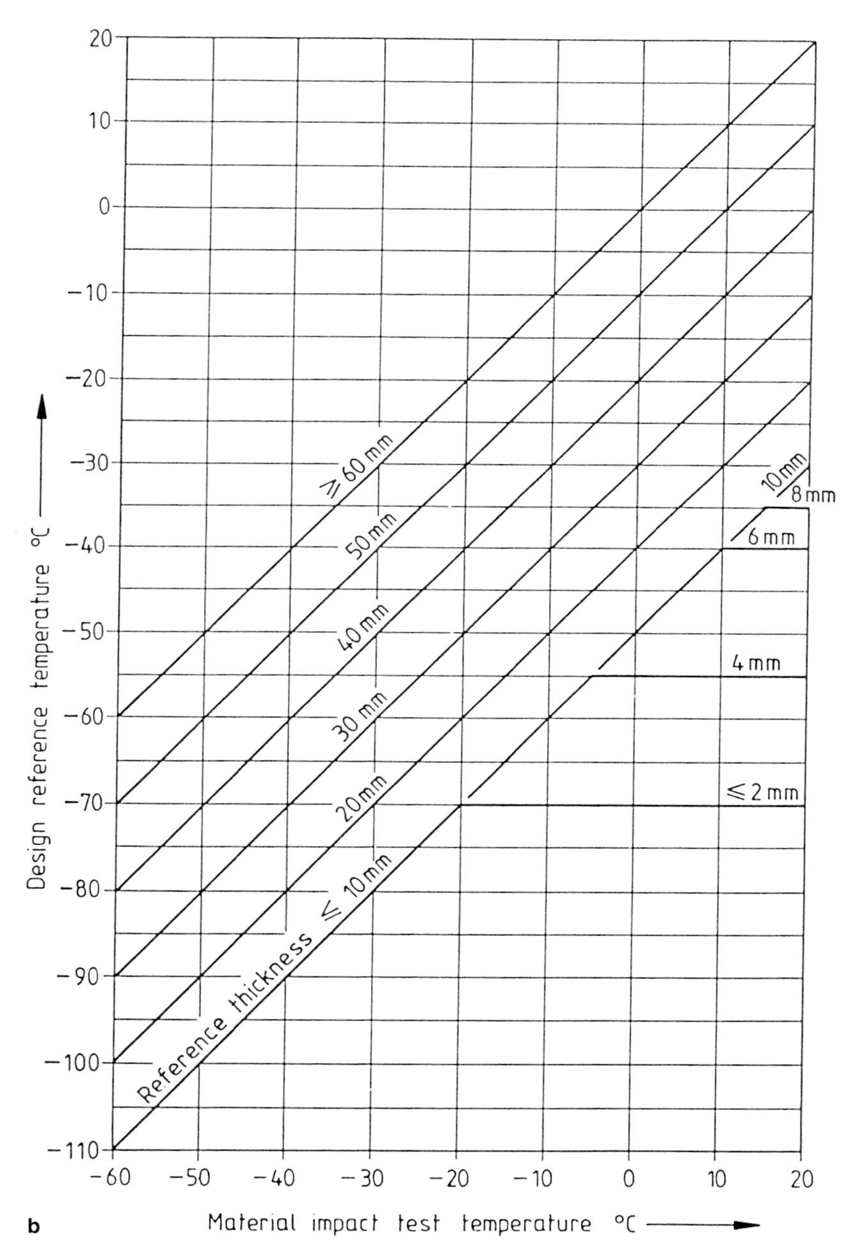

Fig. 3.6 b) Post-weld heat treated components (reference BS5500:1991, please note that this has now been superseded by the 1994 edition).

ageing of strain-aged regions with accompanying restoration of fracture toughness. The effect of this is to permit stress-relieved fabrications in C-Mn steels to operate at considerably lower temperatures than as-welded structures of the same material.[11]

Recommendations for minimum safe operating temperatures for different thicknesses of C-Mn steels with different Charpy V-notch impact properties are given in Fig. 3.6 for both as-welded and stress relieved conditions for applications with static loading.[12] Care must be taken to allow for possible deterioration of properties in service, particularly for steel fabrications operating in the temperature range 200–300 °C.

For particular applications a slow cooling treatment from above Ac_3 may be required to produce an appreciable softening of the material. Alternatively a quenching treatment with a high rate of cooling followed by tempering may be required to give additional strength and toughness after welding. In general, however, these treatments are not applied to welded fabrications of C-Mn steels and will not be considered further.

Low-alloy steels

In the low alloy group of steels, effects of post-weld heat treatment become more complex than for C-Mn steels. The steels themselves are frequently supplied in a heat-treated condition before welding, such as normalised and tempered or quenched and tempered. If post-weld heat treatments are necessary they must be carefully chosen so that optimum properties are developed in the welded joint without destroying the inherent properties of the parent material. In some cases the steel may finally be required in a normalised and tempered condition, but for economic reasons fabrication may be carried out on material in the normalised only condition. Tempering to develop optimum parent material properties may then be combined with a stress relief heat treatment after welding.

The tempering and precipitation effects which can occur in the HAZ of these steels, both during welding and during subsequent heat treatment, mean that the metallurgical effects of post-weld heat treatment must be carefully assessed. The remarks made above for C-Mn steels with regard to stress corrosion and fatigue failure are equally relevant to low-alloy steels. When brittle fracture is considered, however, effects of post-weld heat treatment on fracture toughness in the HAZ and weld metal may not always be beneficial if the heat treatment is incorrectly stipulated.

Thus it is recommended that fracture tests be carried out on specimens notched in the HAZ of a procedure test plate, and given the intended post-weld heat treatment, to assess the possible effects of such heat treatment.

Creep-resisting steels

Most of the above remarks concerning C-Mn and low alloy steels also apply to creep-resisting steels. Post-heat as an extension of preheat becomes increasingly significant as a method of reducing residual hydrogen levels in the joint. Metallurgical effects during stress relief treatments are extremely important for these secondary hardening steels. In the case of vanadium containing steels particularly, a further problem arises which affects the choice of time and temperature for stress relief heat treatments. This is the development of cracking in the fusion boundary region of the heat affected zone during stress relief or during subsequent service at elevated temperatures, Fig. 3.7. Steels containing boron have also been found susceptible.

Hardening occurs when welded joints in these steels are held in a temperature range of 400–600 °C. The susceptibility of a material to reheat cracking can be assessed on the basis of its alloy element content.[13] The carbide forming elements, Cr, Mo, V, Nb, Ti, cause weakening of the grain boundaries through precipitation of secondary carbides. Precipitation of vanadium carbide begins above 450 °C, but relief of residual stresses does not occur until temperatures over 600 °C are reached. Precipitation effects lead to considerable embrittlement, even at high temperatures, and with stresses still present, cracking can develop.[14]

Even when cracking does not develop, the fracture toughness of HAZ regions of weldments in these steels, subjected to stress relief heat treatments below 600 °C, may be extremely poor. A possible solution to these problems is to heat the weldment

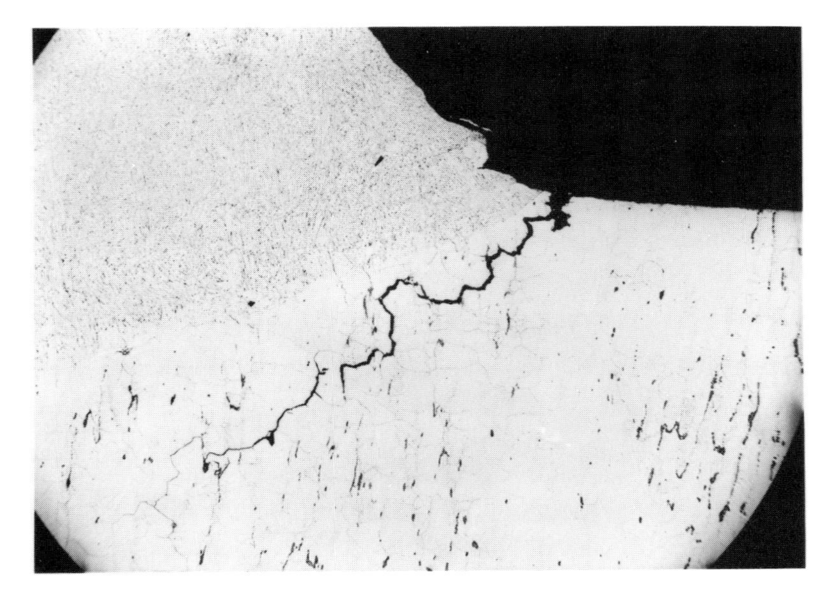

3.7 Reheat cracking in the fusion boundary region of a heat resisting steel.

through the critical temperature range as quickly as possible, so that relief of residual stresses occurs before there is time for precipitation and embrittling effects to cause cracking. Over-ageing of the HAZ regions by heating to 650 °C and above is usually sufficient to prevent the problem arising in subsequent service. This treatment also restores the fracture toughness of embrittled HAZ regions. Care must be taken to ensure that the heat treatment does not impair the properties of the parent material.

In borderline cases it is sometimes beneficial to grind the toes of fillet welds to remove potential initiation sites for this form of cracking and also to use weld metal of lower strength.

In these creep-resisting steels the heat treatment must also be chosen as a compromise to avoid stress relief cracking, to retain adequate toughness in the HAZ and to retain adequate creep strength for service performance. Normalising heat treatments may occasionally be required after welding these steels to assist in developing optimum creep properties in the weld metal and HAZ (see BS2633).

Cr-Mo steels without vanadium are not usually as susceptible to stress relief cracking as those containing vanadium.[15] There are, however, some examples reported of such cracking developing in steels without vanadium in highly restrained joints. Again, the solution is to ensure that the heat treatment brings relief of residual stresses before precipitation effects cause sufficient embrittlement to produce cracking. There is increasing interest in using Cr-Mo steels in the quenched and tempered condition, and, although the steel itself may be so supplied, there is usually no need to repeat the quenching treatment after welding.

In all the low-alloy Cr-Mo and Cr-Mo-V steels the potential benefits of stress relief heat treatments must be offset against the possible disadvantages. Temper embrittlement is a particular problem which may arise in such steels during or after stress relief heat treatment. This phenomenon results in poor fracture toughness of the materials at room temperature due to intergranular weaknesses produced by segregation effects. If susceptible steels are heat treated in the range 450–600 °C, or are permitted to cool very slowly through this range, certain impurities in the steel may segregate to prior austenite grain boundaries and produce an intergranular weakness.[13,16] The elements generally believed to cause the problem are P, Sb, Sn and As (amongst others). It is necessary to use special electron microscopy techniques to detect the segregation effects on fracture surfaces. The embrittlement is found to be completely reversible.

If samples of a susceptible material are embrittled by heating to 550 °C, the inherent toughness can be completely restored by heating above the embrittling temperature range to take the impurities back into solution and cooling sufficiently rapidly to prevent re-segregation to the grain boundaries. Thus, when stress relief heat treatments must be applied to low-alloy steels after welding, the risk of temper embrittlement must be considered and cooling after stress relief should be as fast as is practical.

Heating and cooling rates have to be carefully selected to achieve a balance between being fast enough to prevent embrittlement and slow enough to prevent introduction of undue thermal strains.

Heat treatment should be avoided after welding when it is not necessary to avoid service failures nor to develop optimum properties in a welded joint. The necessity for a heat treatment can be decided only in the light of the particular material, welding conditions, service requirements and risk of failure.

Austenitic steels

Post-weld heat treatment of austenitic steels is necessary to give satisfactory service under some conditions, but may also give rise to other problems. A summary of the situations requiring heat treatment and problems arising from it is given in Table 3.1.

Because of the higher coefficient of thermal expansion of austenitic steel compared with ferritic steel, distortion problems may be more severe in austenitic weldments. Where machining to narrow tolerances has to be carried out after welding, a partial stress relief treatment for example at 550–650 °C or even 400–450 °C, may be required to prevent distortion during machining, as discussed later. Selection of temperatures for post-weld heat treatment of austenitic stainless steel structures for relief of residual stresses is not as straightforward as for ferritic steels, little guidance is given by the codes.

The primary reasons for heat treatment of austenitic stainless steels are to bring about (i) stress relief and (ii) re-solution of harmful precipitates. The extent of stress relief depends on the temperature applied, for example most stress may be removed at above 950 °C whereas around 35% will be relieved at 550–650 °C.[17] Stress relief should be followed by slow cooling but it should be noted that through the temperature range 850–550 °C this will lead to some form of precipitation. Re-solution of precipitates requires temperatures of above 1000 °C followed by rapid cooling, which tends to re-introduce residual stress. Consequently, most stress relief of austenitic steels is a compromise. In addition, use of the highest temperature tends to cause sagging of unsupported structures – a further complication.

The problem of cracking during reheating can occur in certain austenitic steels.[18] It may take place if restrained weldments in susceptible steels are heated, for example to 850 °C for short periods (perhaps less than one hour), or to lower temperatures

Table 3.1 Problems requiring heat treatment of austenitic stainless steels

Material type	Machining stability	Reheat cracking	Transgranular stress corrosion	Intergranular stress corrosion	Weld decay	Sigma precipitation	Knife line attack
304	✓	✓	✓	✓	>0.06%C	✗	✗
304L	✓	✓	✓	✗	✗	✗	✗
316	✓	>0.1%Nb	✓	✓	>0.06%C	✗	✗
316L	✓	>0.1%Nb	✓	✗	✗	✗	✗
321, 347	✓	✓✓	✓	✗	✗	✗	✓
309, 310	✓	✓	✓	✓	✗	Cast,>0.1%C	✗
18/37 (cast)	✓	✗	✓	✓	✗	✗	✗
308, 309 weld metal	✓	✗	✓	✗	✗	✓	✗
Applicable HT	400–450 °C 550–650 °C 850–950 °C 950–1050 °C Slow cool	950–1050 °C Rapid cool	850–1050 °C Slow cool		950–1050 °C Rapid cool	950–1050 °C Rapid cool	950–1050 °C Rapid cool

3.8 Stress corrosion cracking in weldment in austenitic stainless steel.

for longer times, for example to 500 °C for about 1000 hr. Susceptibility to cracking increases with restraint of joints and therefore becomes greater for thick weldments than for thin. All the common austenitic steels have been found susceptible to this form of cracking after welding, except type 316 steel with Nb content <0.1%. Type 347 steel is particularly susceptible. The problem can be prevented by heat treating a weldment at temperatures of 950–1050 °C, to reduce the precipitation which contributes to cracking. Rapid heating through the lower critical temperature regions is required.

The presence of welding residual stresses gives rise to increased risks of stress-corrosion cracking in austenitic weldments in certain environments,[9] see Fig. 3.8. All the common relatively low alloy austenitic stainless steels suffer from transgranular stress-corrosion cracking in chloride or hydroxide environments.

The concentration of chloride ions sufficient to cause cracking may be as low as one part per million in oxygenated water, whereas in the case of hydroxyl ions concentrations of some 40% are necessary. Consequently precautions should be taken during manufacture to eliminate contact with chloride ion bearing substances – even to the extent of controlling the type of solvent used in marker pens. Stress corrosion cracking occurs only in the presence of tensile stress, so that if the total stresses are reduced by relief of residual stresses, susceptibility to cracking may be greatly reduced or eliminated. This calls for a heat treatment to produce relief of the bulk of the residual stresses and requires temperatures of 800–1050 °C followed by slow cooling.

A second form of corrosion cracking can arise in some austenitic steels in marginally/moderately oxidising environments. This form of attack is accelerated by stress and occurs by an intergranular mechanism when chromium-rich $(Fe,Cr)_{23}$ C_6 precipitation has taken place at grain boundaries.[19] This typically occurs during exposure to temperatures in the range 500–850 °C, either in service or during welding. A steel in which this condition has occurred is said to be 'sensitised' to intergranular attack. The sensitivity is a result of chromium depletion adjacent to grain boundaries. It is avoided by carrying out a heat treatment at about 950 °C or above, so that the precipitates are taken back into solid solution, followed by rapid cooling to prevent re-precipitation at the grain boundaries. Care is needed in localised heat treatment of stainless steels, as the temperature gradient either side of the heated band will subject material to temperatures in the critical range for precipitation.

A related problem, which may arise in as-welded austenitic fabrications in marginally/moderately oxidising environments, is that of 'weld decay'. This is a form of preferential intergranular attack in the HAZ of a weld. It is also caused by precipitation of grain boundary chromium-rich carbides in the region of the HAZ which is heated to temperatures between 500 and 850 °C. Susceptible materials should be given a post-weld heat treatment to a temperature greater than 950 °C to take the carbides into solution.

Carbide precipitation problems are largely overcome in stabilised grades, for example types 321 and 347, by addition of elements (Ti and Nb) which form carbides preferentially to chromium. However, in petroleum refinery units, where polythionic acids may form, a stabilisation heat treatment (for example at 850–920 °C for type 347) is sometimes required for resistance to intergranular attack.[19]

A second means of overcoming sensitisation is to reduce carbon levels in the steel. Low carbon austenitic grades (304L, 316L) are available widely with carbon contents equal to or less than 0.03% C, which largely eliminates carbide precipitation related corrosion/stress corrosion problems. In fact, with modern steel making practices, carbon levels in wrought austenitic stainless steels are generally significantly lower than in previous years and problems due to carbide precipitation, especially during welding, are much less common than say 30 years ago.

A further difficulty which may arise as a result of heat treatment after welding is that of 'knife line attack'. This occurs preferentially in the fusion boundary region of weldments in stabilised steels, for example types 321 and 347, after heat treatment in the temperature range 550–750 °C for more than one hour. This attack requires a very specific oxidising environment such as boiling nitric acid and is not a widespread problem. It does not occur if the carbon level is below 0.03%. For susceptible materials and the appropriate environments, heat treatment, if necessary, should be carried out above 750 °C.

During extended periods of service at high temperatures (around 500–850 °C) it is possible for a brittle intermetallic phase to form, known as sigma phase.[20] The times and temperatures at which this happens are different for different grades of austenitic steel, but in general it occurs during service at 650–850 °C for periods from 10 hr in type 304 steel to 1000 hr in type 310 steel.

Sigma phase is generally not present in the HAZ or weld metal of as-welded structures, nor is it likely to occur during brief postweld heat treatment of fully austenitic material, since the times involved are usually too short. However, it may be found during heat treatment of weld metals containing appreciable amounts of retained δ ferrite (for example type 308, 309 weld metals), as formation of sigma from ferrite is considerably more rapid than from austenite. Recent developments with 18Cr–8Ni austenitic stainless steels show that sigma phase formation can be retarded by modifying the chemical composition. Sigma phase will probably have little effect on elevated temperature properties, but can cause marked embrittlement when the weldment has cooled from service to ambient temperatures. Thermal shock due to rapid cooling may cause cracking of embrittled structures.

Similar problems may arise during high temperature service of high carbon cast austenitic materials of the basic 25/20 or 18/37 compositions. In these materials it is possible for extensive precipitation of carbides and sigma phase to occur at temperatures in the range 500–850 °C for times of less than 1 hr which make the material brittle from room temperature up to about 750 °C.

This is most severe with high carbon contents of around 0.4% and becomes less severe with carbon contents of the order of 0.1%. A solution treatment at 1250 °C can be used to take the carbides back into solution to restore the material to its initial condition, but re-precipitation will quickly occur during service in the 500–850 °C range. Wrought materials of similar composition with respect to chromium and nickel generally have a lower carbon content and this problem does not occur. Except where high carbon levels are required for high temperature strength, the most suitable approach to use of austenitic stainless steels where a heat treatment is required is to select one of the low carbon grades.

Mixed austenitic/ferritic weldments

The coefficients of thermal expansion of austenitic and ferritic steels are sufficiently different that, when a joint between them is heated, thermal stresses are produced by the different amounts of expansion. In addition, whilst a temperature can be chosen at which the ferritic steel can be assumed to have relatively low stress levels, it is probable that higher residual stresses will still be present in the austenitic steel at this temperature. On cooling from such a temperature, shrinkage will occur faster in the austenitic steel and, when the joint finally reaches room temperature, appreciable residual stresses will still be present in both components.

Thus, post-weld heat treatments of mixed ferritic/austenitic joints cannot be expected to produce relief of residual stresses. Such heat treatments may be necessary to produce tempering effects in the HAZ of the ferritic part of the weldment, and if this is required the heat treatment must be carefully chosen so that the properties of the austenitic steel are not reduced unacceptably. It is rare that a heat treatment can be selected for a ferritic HAZ without causing some precipitation in an adjacent austenitic weldment, for example made with 308 or 309 type consumables.

One approach is to 'butter' the weld face of the ferritic component with a compatible austenitic welding consumable and post-weld heat treat at this stage. The weld to the austenitic component can then be carried out as if it were a similar joint, without creating a further HAZ in the ferritic steel.

The above remarks also apply to clad pressure vessels in which a thin layer of austenitic steel is applied to the surface of a ferritic vessel for corrosion protection. It is inevitable that residual stress will remain in the region of the austenitic/ferritic interface.

Duplex austenitic-ferritic steels

Heat treatment of duplex stainless steel items for stress relief is not generally applied. This is a consequence of the relatively rapid precipitation of intermetallic phases (for example sigma) and nitrides from the ferrite phase at temperatures in the range 550–1050 °C. Some precipitates significantly reduce corrosion resistance and toughness. The primary reason for heat treatment of duplex steels is to take undesirable precipitates back into solution, i.e. solution annealing.[21] Temperatures of 1050–1250 °C are typically used, followed by rapid quenching, for example with water. Slow cooling must be avoided.

Welded fabrications are rarely solution annealed but, when they are, a matching filler should be used to give a suitable phase balance, rather than the more common overalloyed fillers which are designed to be used without heat treatment.

Lower temperature heat treatments are only applied to dissimilar metal joints to ferritic steel or to cladding when the code being followed demands heat treatment for the ferritic steel HAZ. Choice of a suitable temperature is not well defined, as some loss of properties is expected at all temperatures suitable for the ferritic steel. However, temperatures below 800 °C are tentatively suggested where post-weld heat treatment is unavoidable, which covers most ranges for ferritic steels, that is 550–650 °C.[22] Soak times need to be carefully considered to optimise sensitisation risk and the requirements of the pressure vessel code.

Other heat treatments

In comparison with preheat, post-heat and post-weld heat treatment, normalising of welded structures is a relatively little used practice, the most common application being welded joints in C–Mn steels made using the electroslag process.

The electroslag process, which is particularly attractive for welding thick sections with a single pass weld, brings with it problems of excessive grain coarsening in the overheated regions of the HAZ, close to the fusion boundary. Also, the cast structure of the weld metal itself sometimes has relatively poor fracture properties compared with those in plate material of similar composition. In some cases the fracture toughness of the HAZ and weld metal of electroslag welded joints is regarded as unacceptable in the as-welded condition.

Considerable improvements in the fracture properties of these regions can be achieved by a normalising treatment after welding, since such a treatment produces a considerable reduction in grain size in both weld metal and HAZ. There is some evidence, however, that electroslag welded joints can be used in the as-welded condition if adequate properties are retained in the weld metal and HAZ, or that such joints can be used in the stress relieved condition only, even though there is some deterioration of properties in the joint compared with the parent plate condition.[23,24]

There is a strong argument for accepting the reduced tolerance for defects in zones of poor toughness, provided that increased inspection standards are stipulated to locate defects smaller than would be detrimental in joints made by manual metal arc welding. This may, in some cases, be less costly than normalising treatments. Normalising treatments are not usually necessary for post-weld heat treatment of joints fabricated by processes other than those with a very high heat input. However, modern welding practice relies on submerged-arc welding for joining heavy seams and electroslag welding is much less popular than it was during the 1960s and 1970s. It is worthwhile noting that a knowledge of the normalising and tempering history of the parent material is essential when determining post-weld heat treatment holding temperature ranges. Where a steel has been normalised and subsequently tempered to acquire a given set of mechanical properties, the upper limit of the hold temperature should be set so as not to exceed the tempering temperature. To do so would result in a detrimental reduction in mechanical properties.

At the opposite end of the temperature spectrum for thermal treatments is the process of low temperature stress relief.[25] This method uses temperatures at which metallurgical change will not occur, usually 200 °C maximum. By applying heat rapidly in a controlled manner to either side of the weld a temperature differential is created, and expansion of the parent material imposes a tensile plastic strain on the weld metal. This overloads the stress system in a similar manner to that achieved by application of external loads used for mechanical stress relief. Rapid cooling follows. Because it is difficult to establish the correct heating position and conditions this technique is limited to relatively simple detail designs. A similar technique of heating bands alongside a welded joint has been used by Guan *et al* for controlling residual stresses and distortion in thin welded joints.[26]

An alternative solution

Mechanical stress relief treatments are a process of stress re-distribution during which the high peaks of an existing residual tensile stress are removed by controlled localised yielding.[25] Externally applied loads in the same direction as the residual stress cause yielding in regions where a high stress already exists.

The release of this external load takes place elastically and results in a reduced peak value of residual stress. Although mechanical stress relief can be used to relieve residual stresses produced by welding, there are no accompanying metallurgical benefits as with thermal treatments. They can be effective where a thermal treatment is not practicable, but cannot be considered a complete alternative. Benefits can be gained from the redistribution of local stresses around notches and stress concentrators, providing improved fracture strength at low temperatures and improved fatigue life in these regions. This phenomenon is known as prestressing. However, a comprehensive knowledge of the responses of steels to such treatments is essential, and great care is needed in determining the direction and magnitude of the applied loads.

Chapter 4

Non-compliance in heat treatment practice and potential sources of failure

Non-destructive examination of welded joints and welding itself concern themselves with the immediate joint/fusion face/adjacent parent material relationship. The essence of heat treatment operations necessitates the whole or significant sections of the component parts being raised to a required temperature. The higher the temperature of a heat treatment operation, the more severe are the consequences of a non-compliance. Therefore the following comments are directed at post-weld heat treatment, although the effects of non-achievement in lower temperature operations can be judged proportionately.

Results of non-compliance

Non-achievement of heating and cooling rates

There are virtually no technical consequences of using rates lower than specified, with the exception of certain types of steel whose exposure to elevated temperatures during fabrication and heat treatment has to be critically controlled to guarantee service life. This applies to some low temperature service or high strength steel grades. Apart from these, the consequences are normally of a commercial nature.

The opposite non-compliance, that is, too fast a heating or cooling rate, can affect service integrity by inducing additional stresses in the component. The consequences could be immediate in the form of distortion or cracking, or longer term in reduced service life. If detected, and providing that no permanent damage has been caused necessitating repair, remedial action may be limited to re-heat treatment with the correct cycle.

Failure to reach lower soak temperature limit

The correct post-weld heat treatment temperature ranges are determined by a steel's chemical composition. Hence, lower limit temperatures are generally selected to be high enough to allow relaxation of stresses within a realistic timespan, accompanied by some tempering of the structure. Tempering reduces high hardness and lack of ductility caused by the relatively rapid cooling of the molten weld metal and re-cystallised fusion faces, whilst retaining the basic mechanical properties of the parent material upon which the component design has been based.

Under-achievement of soak temperature results in less tempering of the weld and heat affected zone, and in a reduction in relaxation of stress. Both of these influence the integrity of the component, with the potential to reduce service life if not identified. However, the situation is recoverable, the remedy being re-heat treatment at the correct parameters.

Failure to achieve soak time

The consequences are similar to those above, as is the remedy.

Exceeding upper soak temperature limit

This is the worst non-compliance. The correct upper limit depends on composition but must also take account of the equilibrium diagram for the steel. This identifies the temperatures which are used during manufacture to determine treatments to obtain the required mechanical properties of the steel.

The temperatures of subsequent component manufacturing operations are carefully monitored to avoid reversal of these changes and consequential loss of mechanical properties. Therefore, in post-weld heat treatment the upper temperature is set to give a safe working limit below the lower critical line on the equilibrium diagram. Higher temperatures jeopardise both the welded joint and the parent material properties, affecting the integrity of the structure, and remedial action is no longer confined to the weld itself. Remedial work if required, or indeed possible, becomes costly.

Effects in service

The consequences of omitting a heat treatment or applying one incorrectly could show themselves in a number of ways.

Distortion, machinability and stability

Welding low alloy and other air hardening steels may produce hard zones which make accurate machining difficult. Correct heat treatment softens the hardened zone sufficiently to make machining satisfactory, and sometimes the primary benefit is softening of the material rather than elimination of residual stresses.

Structures which must maintain a high degree of dimensional stability frequently require stress relieving before machining. Residual stresses caused by welding form a balanced system within the structure and machining may unbalance the system by removing the material containing these stresses (Fig. 4.1). Consequently distortion may occur during machining or in service. An example of this could be a simple ring welded by a number of butt joints. When the surface layers of the ring are removed the balance of residual stress is upset and warping readily occurs. When there is doubt as to whether stress relief is required, heat treatment is often specified as a safety precaution. Indeed there could eventually be a financial saving.

This problem is common to all materials, but is more severe in austenitic steels and non-ferrous alloys than ferritic steels.

Initial cracks

Various forms of cracking can arise as a result of fabrication by welding. Cracks can develop in the weld metal or the heat

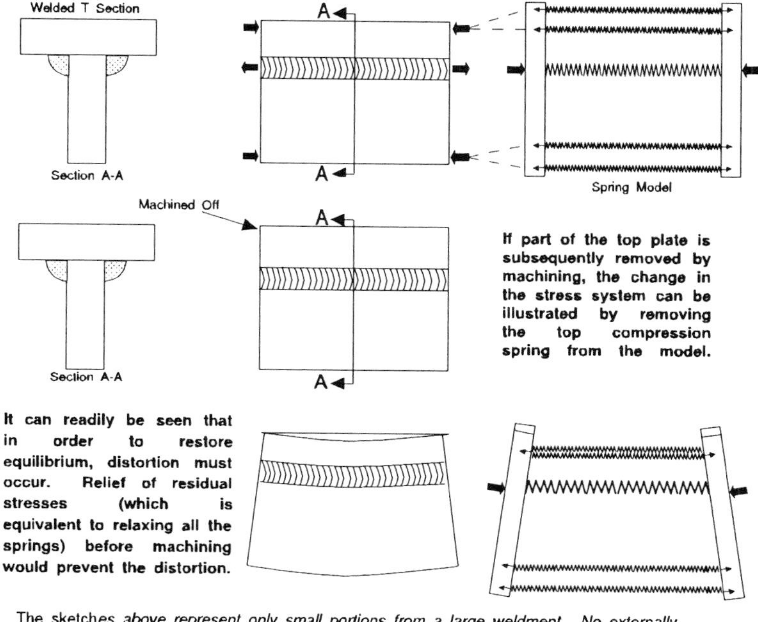

4.1 Distortion caused by residual stress in a machined component.

affected zone. Most of them occur during cooling of the weld from high temperatures, so that they are already present when the weld is completed. The main exception to this is hydrogen cracking. Heat treatment after welding is of no help in preventing welding cracks if they have already occurred on cooling.

To prevent hydrogen cracking it is sometimes necessary to maintain a preheat and follow it immediately with a stress-relief/tempering treatment to soften the heat affected zone and make it less susceptible to cracking at room temperature. This is the only case in which post-weld heat treatment need be considered as a means of preventing welding cracks. Unless it can be carried out immediately after welding, that is before the whole weldment has cooled to room temperature, there is no guarantee that the heat treatment will prevent HAZ cracking.

Apart from cracks which may occur during welding, the possibility of them developing during heat treatment must be considered. This form of cracking can occur in vanadium and boron-containing ferritic steels and in some austenitic steels. Where heat treatment is essential for various reasons, or where high temperature service is involved, precautions must be taken to prevent reheat treatment cracking. These precautions usually consist of raising the temperature through the dangerous precipitation range as quickly as possible and reducing stress concentration at potential initiation sites by grinding the toes of fillet welds. It is common practice with pipe butt welds in Cr-Mo-V steels to maintain preheat whilst the cap of the weld is blended by grinding, followed by immediate post-weld heat treatment (see BS2633:1987, Para 21). Use of softer weld metal can also alleviate the problem in borderline cases.

Corrosion : caustic embrittlement

When mild steel is to be exposed to a corrosive environment the surface is usually protected by paint or other coatings, and so stress relieving significantly does not usually improve the resistance of mild steel to corrosion.

Pressure vessels, boilers and other fabrications which are exposed to caustic solutions frequently develop caustic embrittlement and stress corrosion cracking at local areas of high stress. Stress relieving can reduce the residual stresses caused by welding and can prevent caustic embrittlement. The treatment is therefore definitely beneficial.

Unstabilised austenitic corrosion resistant steel (stainless steel) may be damaged when subject to a stress relief heat treatment. This is because stress relieving temperatures in the range 550–850 °C sensitise the steel and make the entire surface vulnerable to corrosive attack. This does not generally apply where the low carbon grade of an austenitic stainless steel is involved. Clearly, to reach the higher temperature ranges needed for effective stress relief, namely 800–850 °C, it is necessary to pass through the sensitising temperature range on both heating and cooling. To avoid damage, cooling should not be retarded through the 850–500 °C range.

A great deal of thermal heat treatment is carried out on petrochemical plant and construction sites, particularly on pipework being erected in the field. This pipework is mainly Cr-Mo but also includes 9% and 12% Cr materials. To prove the effectiveness of the heat treatment a maximum Brinell hardness figure is usually specified. This ensures that the necessary mechanical properties are retained in the weld metal, heat affected zone and parent metal after heat treatment.

Stress corrosion cracking is usually caused by the fluid carried by the piping, mainly solutions of caustic alkalis, some nitrates particularly when hot, and liquor containing HCN or cyanogen compounds.

Application of stress-relief treatments after welding to avoid potential stress corrosion is of considerable benefit. This arises partly from removal of residual stresses and also from tempering effects in the HAZ of ferritic steels. Where stress corrosion is known to be a risk in the projected materials and environments, stress-relief heat treatment may be the only way to obtain satisfactory performance in service. The heat treatment temperature, however, need not necessarily be high enough to eliminate residual stresses completely; much lower temperatures than the usual 600–650 °C for mild steel are frequently adequate.

Stable crack growth

The effects of post-weld heat treatment on fatigue have been mentioned previously. In general, effects of stress-relief treatments are small when part of the fatigue loading cycle is tensile. Some benefit can be derived, however, when the fatigue cycle is mainly compressive. Considerable improvements in fatigue performance can be gained by local alterations to residual stress distribution, either by spot heating or by peening, to introduce compressive stresses at potential initiation sites.

Final failure

In pressurised plant, failure has occurred when cracks develop to such an extent that they penetrate the wall thickness and leakage takes place. This is by no means as serious as complete destruction of a fabrication by fast fracture, although great inconvenience may be caused.

In many materials considerable benefits in resistance to failure by brittle fracture can be obtained by post-weld heat treatment. In C-Mn steels this arises partly by reduction of residual stresses and partly by removal of local embrittlement effects due to welding. Heat treatment does not affect resistance to fracture propagation in C-Mn steels and use of stress relieved fabrications at low temperatures relies upon prevention of fracture initiation.

In low-alloy ferritic steels the precipitation effects which may occur during stress relief heat treatments should be carefully considered. In some materials it is possible for fracture toughness to be impaired by incorrect heat treatments, even though some benefit is gained by removal of residual stresses. Improvements through tempering of hard zones may not be obtained if the temperature is not controlled properly. Nevertheless, if some degree of embrittlement occurs in weldments it is usually possible to select post-weld heat treatments to achieve some improvement. The temperatures required for satisfactory performance are much more critical with alloy steels than for C-Mn steels.

In austenitic steels brittle fracture is not usually a problem and heat treatments to assist in preventing fracture are not necessary.

Chapter 5

Methods of heat treatment

Whether hydrogen bake-out, preheat, post-heat or post-weld heat treatment is being considered, the methods used can be broadly categorised in terms of energy source – electricity or combustible fuels. The method selected will depend mainly on the size, geometry and material of the fabrication being welded, as will the type of heat treatment. The objective of any heat treatment operation is to achieve required temperatures uniformly or within specified limits on gradients, in a controllable manner, all to a specification which will enhance the integrity of the structure. As with all industrial processes, due regard must be paid to the commercial pressures which demand the most economical solution.

Preheating

By their nature, preheat operations are carried out using portable equipment which can be easily applied and removed from the welded joint. The objective is to heat an area of parent material either side of the weld preparation, typically through the thickness and up to 150 mm from the edge of the fusion face, and maintain the temperature throughout the welding process. Several methods are available and their selection may well have to reflect the length of weld and overall geometry and the need for concurrent post-heat operations, as well as commercial considerations. Ideally heat should be applied to the opposite surface to that on which welding will be carried out and a steady temperature achieved at least 50 mm and preferably 75 mm from the weld

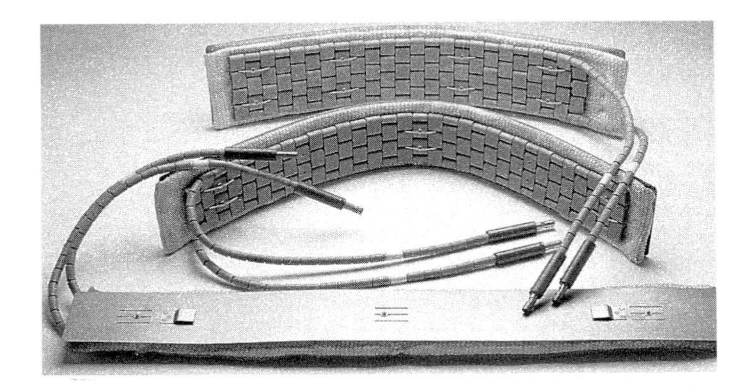

5.1 Low voltage flexible preheaters (courtesy Cooperheat (UK) Ltd).

position to ensure that the required preheat has really been attained. This may not always be practicable, and often heat is applied from the same side as the welding operation.

In its most basic form preheating can be carried out using a gas flame, the flame being played over the joint until the required temperature has been reached, as determined using temperature indicating crayons. This method relies heavily on the operator to ensure its correct application. Ideally the preheat temperature should be checked on the side remote from the gas flame. However, when this is not practicable, measurements should be made some time after removal of the heat source to avoid misleading results because of high surface temperatures. A delay of one minute for each 25 mm thickness of parent metal is suggested.

Gas flame preheating is frequently used for smaller fabrications and pipe sizes in carbon and carbon-manganese steels where more advanced methods would not only impair welder access, but also may prove unnecessarily expensive. Generally speaking, the duration of welding on such small joints relative to the time taken to install preheat equipment, and the subsequential heat input from the process itself, makes this approach adequate, if not the optimum in controllability.

As the joint size and/or the complexity of the steel increase, methods have to be applied which use equipment attached to

5.2 Application of flexible preheaters.

either side of the joint. The equipment must maintain the preheat temperature uniformly along the total length over a time sufficient for welding to be carried out.

Perhaps the most widely used method of controlled preheating is to use low voltage contact electrical resistance heating elements. These heaters are manufactured by threading stranded

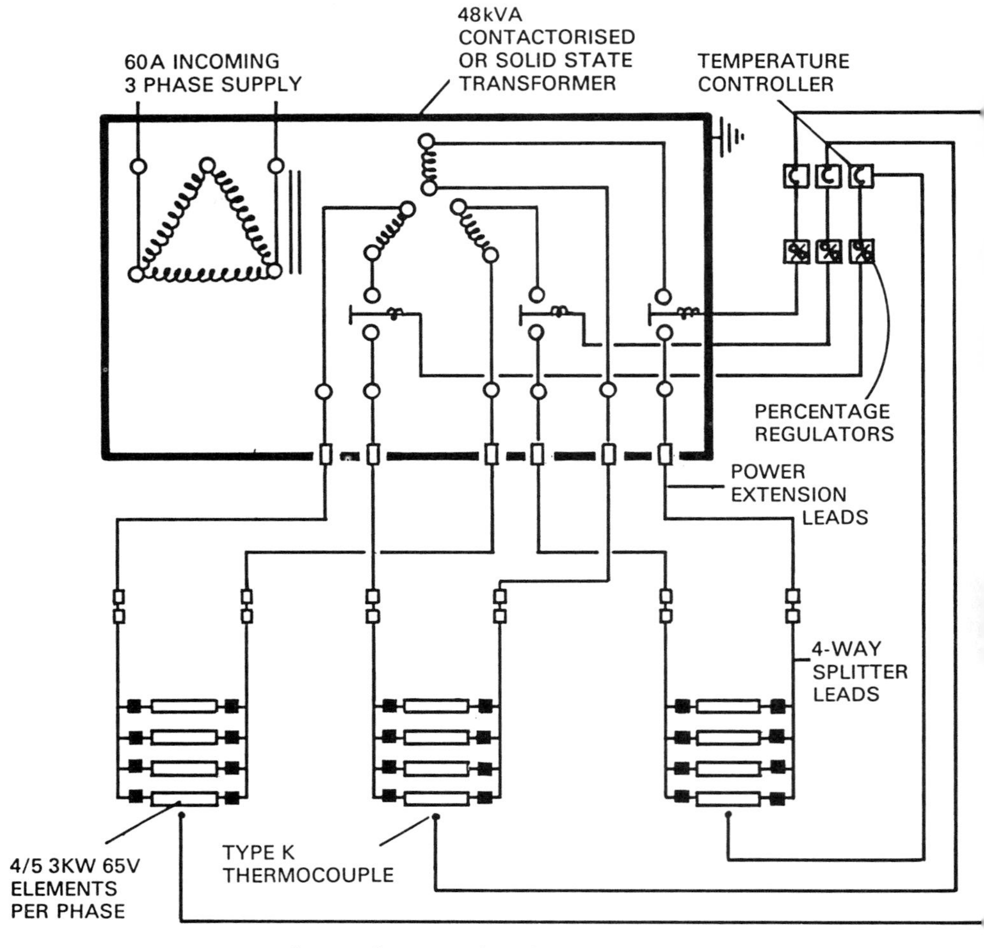

5.3 A low voltage preheating system.

resistance wire made from 80/20 nickel-chromium or 60/15/25 nickel-chromium-iron into interlocking sintered alumina ceramic beads. This principle enables heating elements to be manufactured in a range of sizes, within the limitation of the length of wire required to operate at the given low voltage. In a pre-insulated modular form (Fig. 5.1) these elements are ideal for preheat applications and can be held in contact with the workpiece by either banding or magnet pairs (Fig. 5.2).

5.4 Preheat by surface combustion radiant gas heaters (courtesy Mitsui Babcock Energy Ltd).

A number of elements can be connected together to form a zone, each zone being controlled by a thermocouple. This would usually be sited 50 mm from the edge of the weld preparation so as not to impede welder access and to protect the hot junction from spatter. The low working voltage can be provided either by welding machines of 300/400 amp capacity or more generally by custom-built heat treatment power sources, which can be used for a complete range of heat treatment operations carried out locally to the weld.

A typical connection diagram for a zoned low voltage preheat system is shown in Fig. 5.3. A low voltage contact electrical resistance heating system can be applied over a wide range of geometries, from butt welds to large node and plate fabrications. In these the structure remains static and welding is carried out positionally. For some welding applications, typically submerged-arc, the joint is rotated so that welding is carried out downhand. It is possible to use contact electrical resistance heating in such circumstances, but great care is needed to ensure that the power cables do not become entangled, usually by reversing direction after every complete revolution. Where rotation of the component is required, preheating can be achieved using heat transfer by radiation. For this application the heating system is built into supporting steelwork which surrounds, usually, the lower half of the vessel.

Although electrical resistance heating elements are widely used, a more common method utilises surface combustion radiant gas heaters. Burning natural gas, propane or butane, they provide a very effective method of maintaining preheat temperature during the extended welding time needed to fabricate thick walled pressure vessels (Fig. 5.4). These surface combustion units (SCUs) can also be used as an alternative to electrical resistance heating on static structures, especially on construction sites or in fabrication yards where electrical power is at a premium.

Hydrogen bake-out and hydrogen release (post-heat)

As both of these processes require full control over temperature and time at temperature to be effective, controlled techniques which incorporate thermocouples are used. For hydrogen release heat treatments which follow on from preheat, uncontrolled methods such as random flame monitored by temperature indicating crayons are virtually excluded. In hydrogen bake-out, some of the techniques discussed later with regard to local in-situ post-weld heat treatment are used, especially on large structures such as pressure vessels and/or where temperatures of 400–450 °C are required.

Post-weld heat treatment

Wherever practicable, post-weld heat treatment is carried out in a permanent furnace. Furnace construction has changed with the

5.5 A bogie hearth furnace: a) Interior.

advent of the low thermal mass principle of construction and high velocity burners, with consequential improvements in performance, heat-up times, temperature uniformity and fuel economy. The older types of brick furnace can be modified to use high velocity burners and computerised management systems, such as the bogie hearth furnace shown in Fig. 5.5.

The low thermal mass approach, so called because of the rela-

5.5 b) Gas train (courtesy Mitsui Babcock Energy Ltd).

tively low levels of energy absorbed by ceramic fibre and mineral wool types of insulation compared with traditional brick linings, offers immense versatility to the furnace designer. Large top hat type furnaces can be constructed for pressure vessel work, or for heat treatment of steels after hot or cold forming (Fig. 5.6). The principle lends itself to construction of custom-built furnaces to

5.6 A low thermal mass top hat furnace (courtesy Mitsui Babcock Energy Ltd).

accommodate a particular configuration for heat treatment, for example header boxes on pendant elements (Fig. 5.7).

Dependent on size, location, and load configuration both electrical and fuel sources of energy can be used equally well.

The modular construction of low thermal mass furnaces offers a further advantage over traditional designs, in that they can be more readily dismantled and moved within the factory or even to

5.7 Partial heat treatment in temporary furnaces (courtesy Mitsui Babcock Energy Ltd).

another site altogether. Reconstruction times are short without a need for refractory specialists to re-install brickwork.

Use of permanent furnaces offers a number of advantages:

(i) The temperature distribution within a good modern furnace will normally meet the requirements of current fabrication codes or customer specifications.

(ii) By careful consideration of the heating cycle with respect to geometric complexity and section thickness, uniform heating of a complete structure avoids inducing thermal stresses, as is the case with local or partial heat treatment.

(iii) Where operation is automated, labour utilisation can be maximised with cost benefits.

(iv) Once a furnace has been built, effective operational planning can justify the considerable capital investment over several years of operating life, providing a cost effective and economical method of heat treatment.

(v) Operational planning minimises delays to the production programme.

There are, however, many circumstances where a permanent furnace cannot be used for heat treatment. It may be that a final fabrication is too large for the available furnace capacity, or that a structure has to be shipped piecemeal and erected on a site where there are no furnace facilities. In such cases one solution is to heat treat a part or the whole of the structure in a temporary furnace. Using a lighter form of casing construction, such as weld mesh and angle sections (Fig. 5.8), or pre-folded mild steel panels, the low thermal mass principle can be used to great advantage. The insulation lining can be either mineral wool and/or ceramic fibre dependent on peak temperature and expected number of cycles required.

Again, selection of energy source is dependent on a number of factors. For loads weighing over 150 tonnes, fuel firing has advantages in that electrical heating requires large power sources or extensive mobile generator packages.

Where vessels are erected or repaired by welding on site, or where closing welds in pressure vessels join two or more sections such that the final entity cannot be heat treated in a furnace, use of a temporary furnace can be developed into the concept of internal heating. Here, by insulating the outside surfaces and heating from the inside either by electrical power or fuel firing, the vessel or structure becomes its own furnace. This is a very effective technique and is used extensively in construction, repair and modification of large steam drums, petrochemical reactors and storage spheres and tanks.

It is possible to post-weld heat treat a complete vessel using this

5.8 Temporary low thermal mass furnace (courtesy of Mitsui Babcock Energy Ltd).

technique (Fig. 5.9), or to treat part of a vessel by defining the hot zone with temporary fully insulated internal bulkheads. The extent of the hot zone for a localised approach has to meet minimum criteria specified by code requirements. For example, ASME VIII and BS5500 for unfired pressure vessels define the minimum width of heated band and limitations on temperature gradients away from these hot zones. Where nozzles or other load bearing

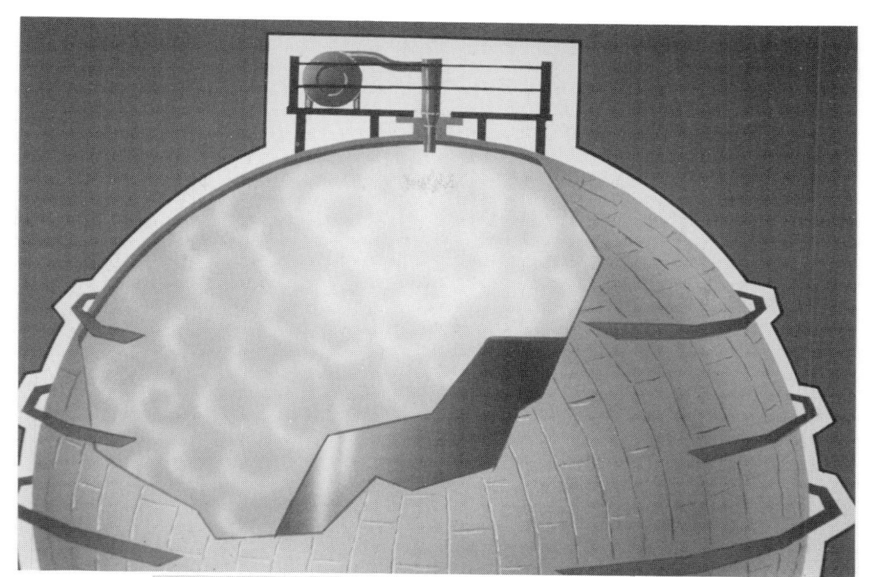

5.9 Post-weld heat treatment of a storage sphere by direct internal firing (courtesy Cooperheat (UK) Ltd).

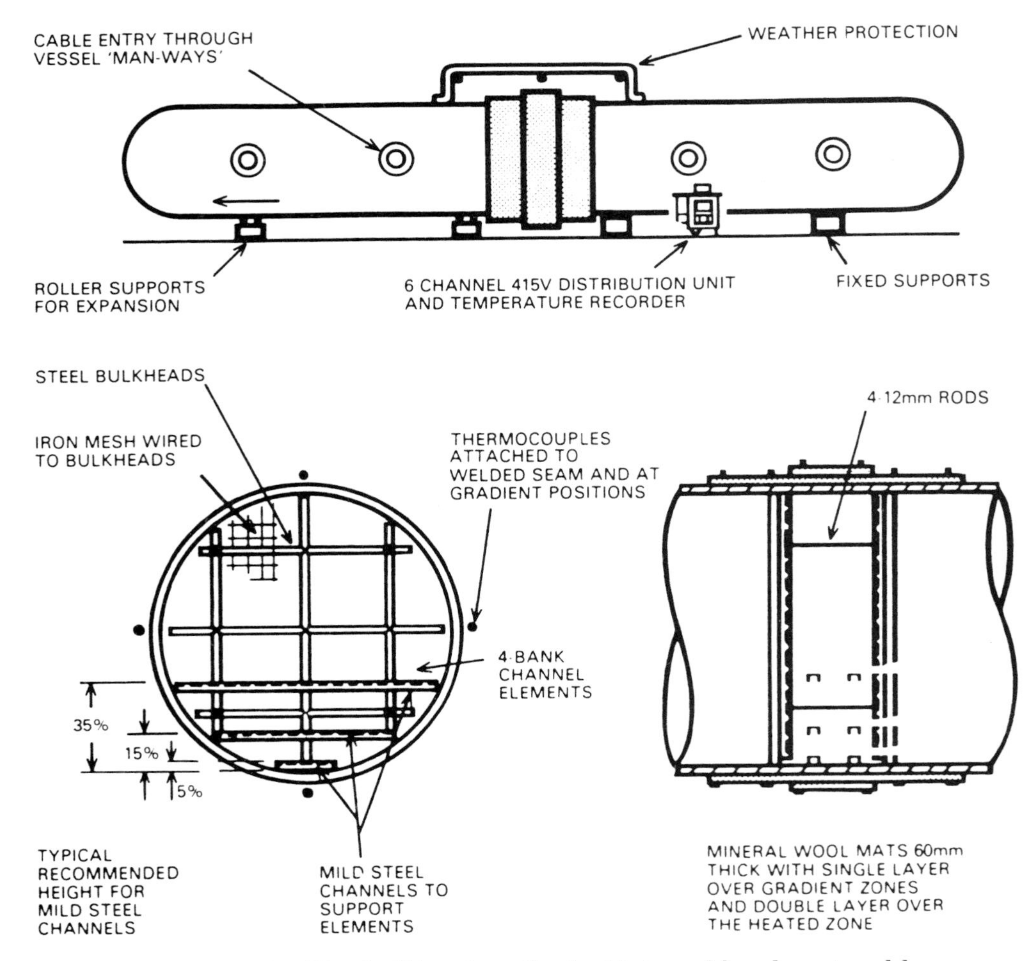

5.10 The bulkhead method of internal local post-weld heat treatment.

attachment welds interfere with the heated zone, or are subject to post-weld heat treatment themselves, the heated band width has to be extended.

Figure 5.10 shows the general principles which apply to post-weld heat treatment of a closing seam on a pressure vessel, using the internal bulkhead method and electrical resistance heating. Figure 5.11 illustrates the temporary steelwork used to support

5.11 Temporary steelwork is used to support the resistance heating elements in the bulkhead method (courtesy Cooperheat (UK) Ltd).

the resistance heating elements. This is a relatively simple case. Where branches lie within the heated area or adjacent to it, or where the heated zone is large, other aspects of the heat treatment procedure must be considered. It is important to ensure adequate support at temperature and freedom to expand. It may be necessary to apply compensating heating systems to ensure that attachment welds reach temperature, to aid temperature uniformity and to control temperature gradients. All these issues should be addressed where local in situ heat treatment of pressure vessels is carried out, to protect the integrity of the structure.

In any large post-weld heat treatment project it is also necessary to assess the stability of the structure whilst at soak temperature. In design the grade of steel selected reflects, amongst other requirements, the operating temperature. Material properties at the design temperature are obtained from the appropriate code. However, post-weld heat treatment temperatures often exceed the highest values of design temperature quoted and consequently material properties are much reduced. Therefore the stability of the structure during heat treatment has to be reviewed and judgements made on support. It may be, as in the case of a horizontal vessel with permanent saddles, that support is adequate. Alternatively the span may be too great and additional temporary saddles are required. For vertical columns heat treated horizontally, support of the whole structure on temporary saddles has to be evaluated. For vertical columns heat treated in situ the question of stability at soak temperature becomes a function of factors such as self-weight and bending due to windage. It is necessary to establish criteria for assessment. A limited amount of quantitative data exists from research, but generally the approach is empirical, extrapolating the design data to the higher temperatures. In view of the lack of precise information, it is advisable to be conservative with any assumptions, and the experienced heat treatment engineer places reliance on past performance of a given method of assessment.

Effects of expansion (and contraction) also require consideration. A ferritic steel expands by about 2.5 mm per 300 mm, at 600 °C. For large fabrications this movement can be significant and constraint must be eliminated where possible.

In treating horizontal vessels, supports should be placed on rollers or as a minimum on greased plates, provided that a high temperature quality lubricant is used. It may be necessary to provide some mechanical aids to ensure that movement takes place, measured against a prior calculation.

Where vessels are being heat treated in situ, as would be the case after repair or modification of existing plant, all ancillary steelwork such as supports, interconnecting pipework, platforms, ladders, will have to be inspected. Where these features restrict free expansion, it will be necessary to disconnect, or at least loosen, the appropriate parts. Any scaffolding surrounding large structures requiring heat treatment in situ should be free standing with sufficient clearance to accommodate these movements.

Irrespective of whether a furnace heat treatment, temporary or permanent, or an internally applied heat treatment is used, it is important to review attachments to the main structure, whether they are incorporated in the pressure envelope or are load bearing, such as nozzles, manways, lifting lugs, support rings, diaphragms and tubesheets. These appendages must be heated in unison with the component to which they are attached, achieving a temperature within the defined soak range and not giving rise to adverse temperature gradients. It may be that to achieve this a supplementary heating system is required to counteract heat losses.

Within some pressure vessel codes, reference is made to control of temperature gradients away from localised heated bands, from definitive guidance to a general statement that a 'temperature gradient is not harmful'. For simple cylindrical shapes, the bending stresses created by localised heating and the associated temperature gradient can be calculated.

For more complex geometries, finite element analytical software packages for personal computers offer a powerful method of investigating effects of localised heating and associated stress distributions. Using this technique the heat balances can be adjusted to produce temperature profiles which offer acceptable stress distributions. This information can then be used to derive a heat treatment procedure incorporating the necessary controls on temperature gradient or any other aspect necessary to provide an acceptable result.

Figure 5.12 shows an example of a butt weld to a nozzle in a

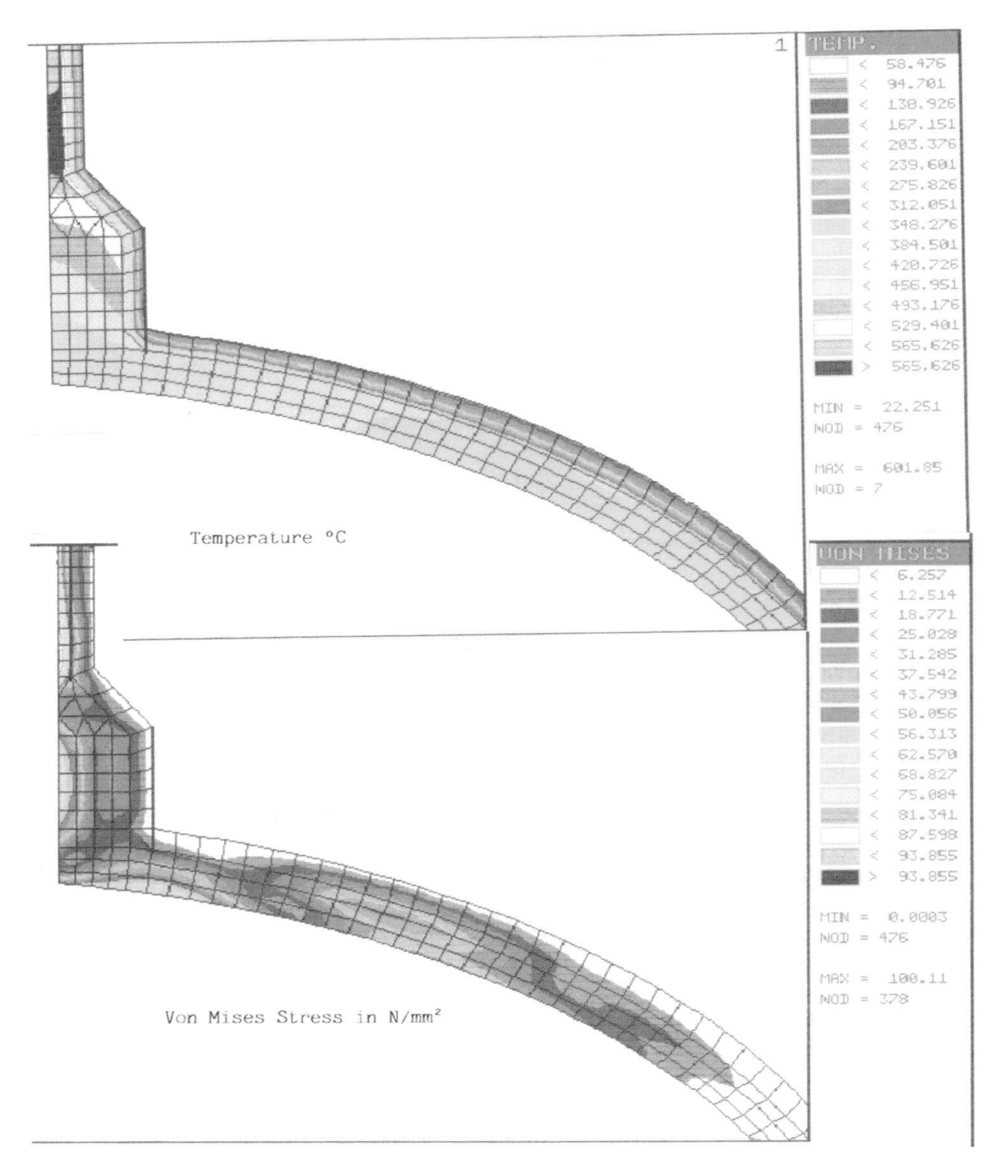

5.12 Temperature and stress analysis for a nozzle to dished head butt weld with background heating (courtesy Cooperheat (UK) Ltd).

dished head. It illustrates the need to control the temperature gradient within the head itself to avoid unacceptably high levels of thermally induced stress being generated in the nozzle to head weld. This is a simple axisymmetric two-dimensional model. Larger programmes are capable of generating three dimensional models.

Other circumstances where furnace heat treatment cannot be used are:

(i) Erection of pipework on engineering construction sites such as power plant, petrochemical plant, offshore platform/ module hook up stages;
(ii) Repair and maintenance, overhaul and upgrading operations on existing plants.

These may involve relatively straightforward pipework joints, or complex configurations such as welds to valves, or nozzles to pressure vessels. Where heat treatment is necessary, it has to be carried out locally using specialised techniques.[27]

Localised heat treatment of pipework butt welds is generally governed by the same codes of practice and standards as applicable to furnace treatments. However, since heat is only applied over a specific distance either side of the welded joint, careful consideration of the techniques used is required. It is important to ensure that the temperature distribution meets the given criteria and that the inherent induced thermal stresses due to temperature gradients away from the hot zone are tolerable.

Tolerance on soak temperature can typically be ±20 °C. Some clients' specifications call for closer tolerances. This condition should normally be achieved at the weld and immediate environs, to include the full weld width plus the heat affected zone. It should extend, say, to a distance of 1.5 × wall thickness each side of the weld centreline, measured on the side from which the heat is applied.

Furthermore, a temperature gradient inevitably develops through the weld thickness. This is because such localised treatments are effected by application of an external source (it should be noted that induction techniques have limited penetration), and

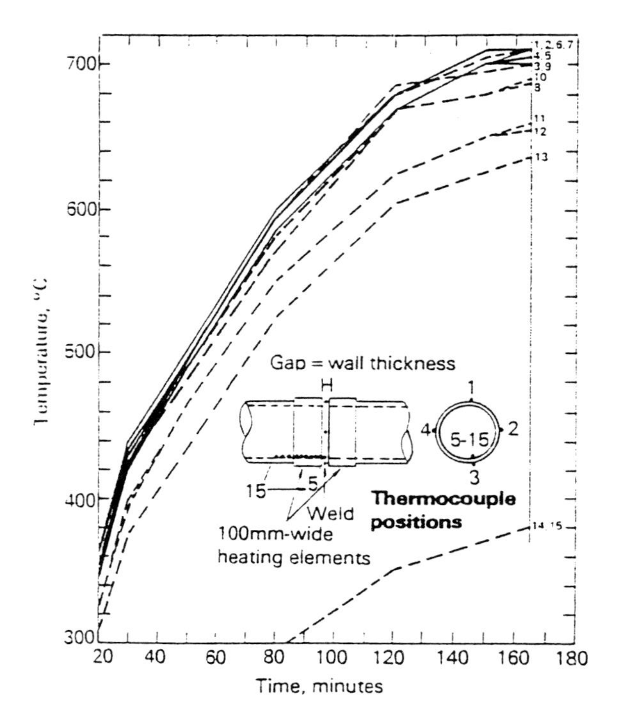

5.13 A through-thickness temperature gradient is revealed by the internal and external temperature profiles in the region of a locally heat treated butt weld (horizontal pipework, 25 mm wall × 240 mm diameter).

also in view of significant internal heat losses by convection and radiation away from the heated band.

It has been demonstrated that to achieve satisfactory through-thickness temperature gradients, a minimum heated band width of five times the wall thickness should be applied,[28] although factors of six or seven are generally used. For a typical heat treatment of a 50 mm section, a through-thickness temperature gradient of about 15 °C is not unrealistic (Fig. 5.13). It is therefore apparent that diameter and thickness should be taken into account in estimating through-thickness gradients, and the heating system adjusted accordingly. Given a 15 °C through-thickness

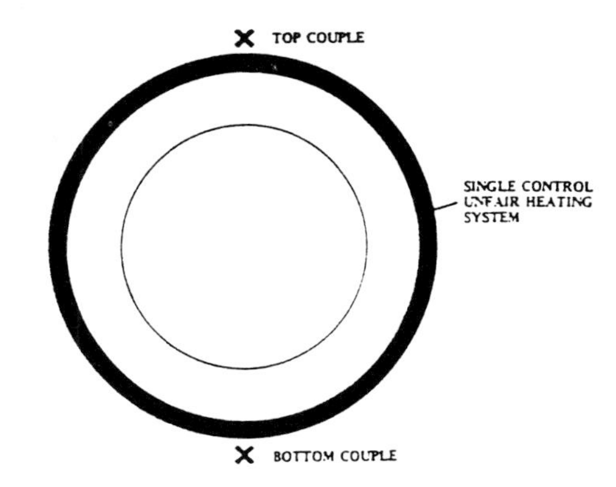

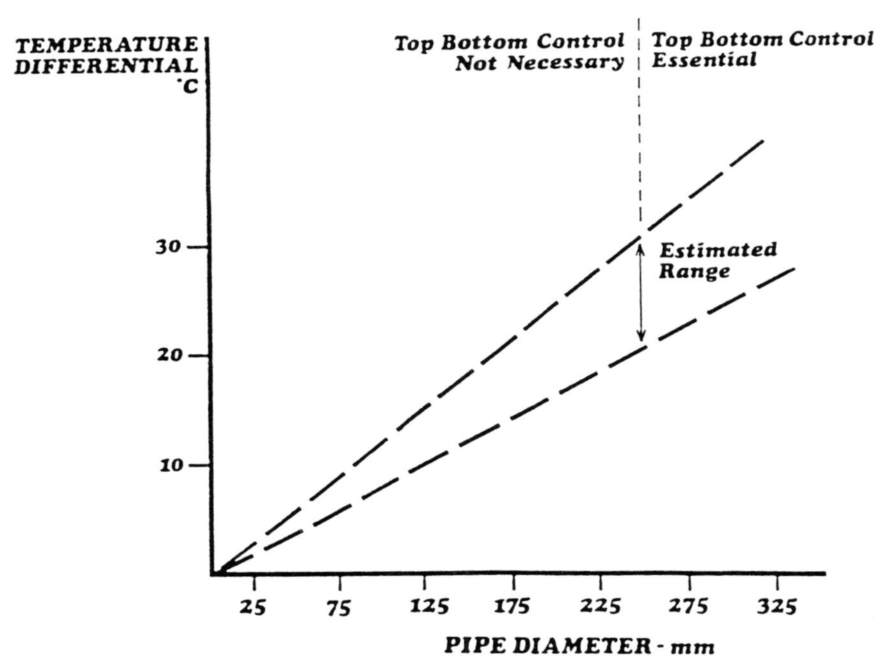

5.14 Top to bottom temperature differential in pipe butt welds.

5.15 Arrangement of flexible ceramic pad heaters for local *in situ* post-weld heat treatment (courtesy Cooperheat (UK) Ltd).

temperature differential, it is important to raise the external surface of the butt weld to as high a temperature as possible, within the specified permitted range, to obtain satisfactory results at the bore.

Where a uniformly-heated band is applied to a pipe butt weld mounted horizontally, a temperature differential arises between the bottom and top weld positions, owing to convection both inside the pipe and externally in the annulus formed by the heating system. This effect is more marked in large pipe sizes where cross-radiation is less significant (Fig. 5.14), and reaches some 25 °C temperature difference for 250 mm diameter pipework.

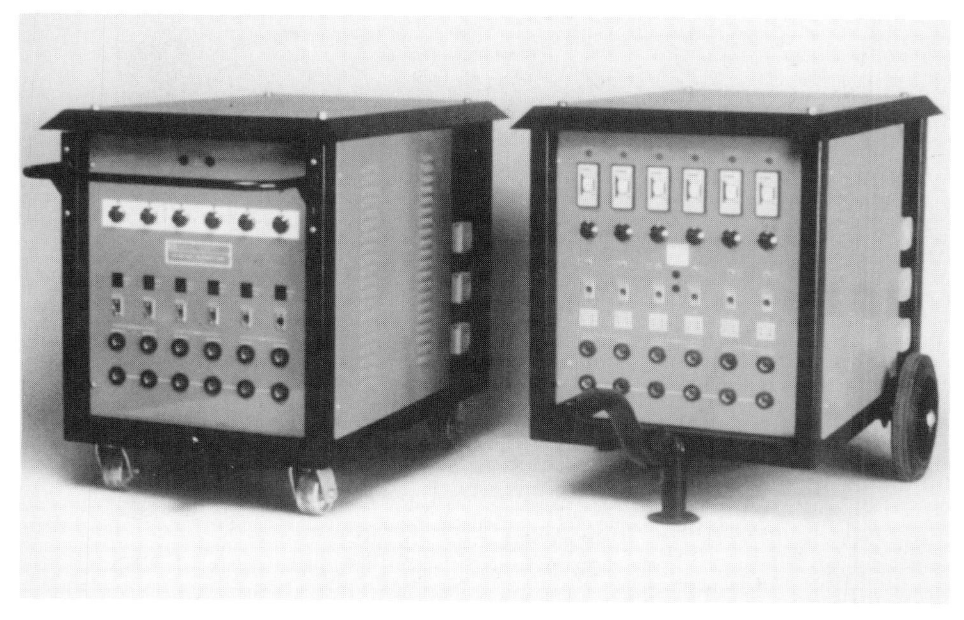

5.16 Low voltage (50 kVA) heat treatment power sources (courtesy Cooperheat (UK) Ltd).

A heating system with a facility for zonal control is advantageous under these circumstances. Use of modular ceramic pad or braided electrical resistance elements can overcome the inherent disadvantage of other techniques using gas burners or induction. Again, use of 80/20 nickel-chromium and 60/15/25 nickel-chromium-iron stranded wires, in conjunction with high quality sintered alumina ceramic beads, ensures long life and a low risk of short circuit on to the workpiece.

Individual elements are normally rated for 80, 60 or 42 V operation, to suit the power source voltage or particular mandatory requirement, and are available in a range of standard sizes. They may therefore be applied over a wide range of pipe sizes and component geometries without recourse to manufacture of a special element for the application (Fig. 5.15). Because of the fixed voltage rating, elements can be parallel-connected in groups to form heating zones, independently controlled, with power provided from custom built heat treatment power sources (Fig. 5.16).

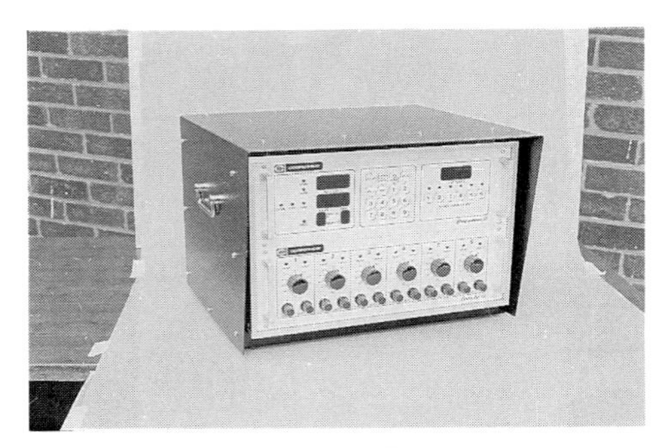

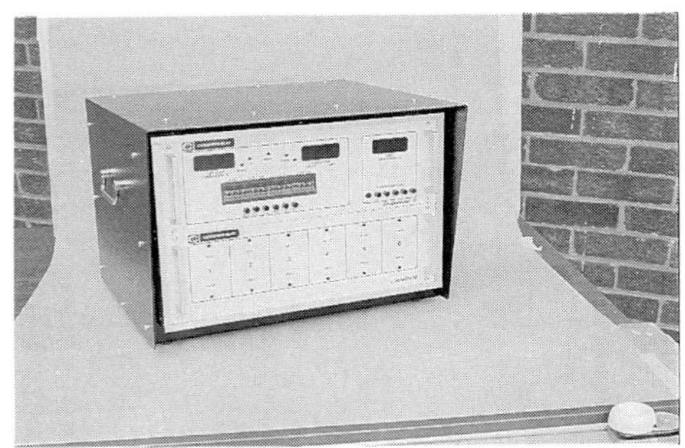

5.17 Heat treatment programmers (courtesy Cooperheat (UK) Ltd).

Automatic control over the full heat treatment cycle can be achieved by programmers which form part of an integrated on site heat treatment system (Fig. 5.17). For small pipework diameters, up to 75 mm, a large element, 60 V, 2700 W, is impractical, giving an unnecessarily wide heated band. Smaller purpose-designed elements are more suitable and are operated from equipment modified to have lower voltage outputs.

Zonal control around the weld circumference is, of course, not necessary where a pipe is vertical. However, power requirements are higher (owing to higher internal convection losses) and, where the heated band is applied symmetrically about the weld centreline, the peak temperature is displaced upwards, resulting in a non-symmetrical profile.

It is therefore normal practice to apply the pads slightly displaced below the weld level or, for heavy sections, to control as two discrete bands, above and below the weld. To provide a means for control, thermocouples are fitted at $1\frac{1}{2} \times$ thickness from the weld centreline. Gaps between pads should be no greater than the wall thickness (Fig. 5.18), so as to avoid excessive variation in temperature and possible overheating of the parent metal.

A further concern in many (but not all) standards is control of longitudinal temperature gradients along the pipe, on each side of the weld. The purpose of the restrictions is to limit the thermal stresses generated by the sharp change in gradient at the edge of the heated band, see Fig. 5.18.

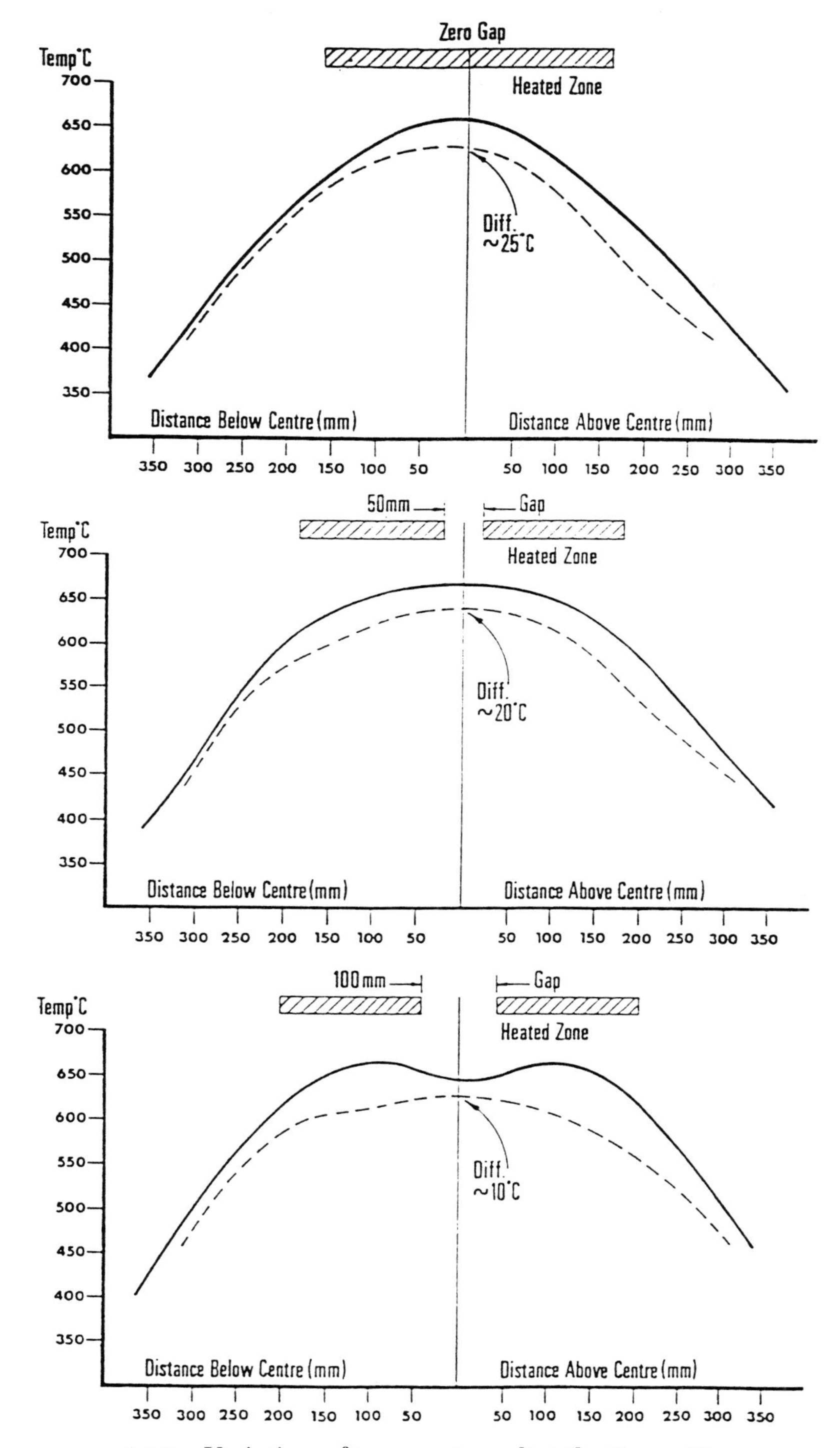

5.18 Variation of temperature distribution with centreline gap for vertical pipework butt welds (solid line, external; broken line, internal).

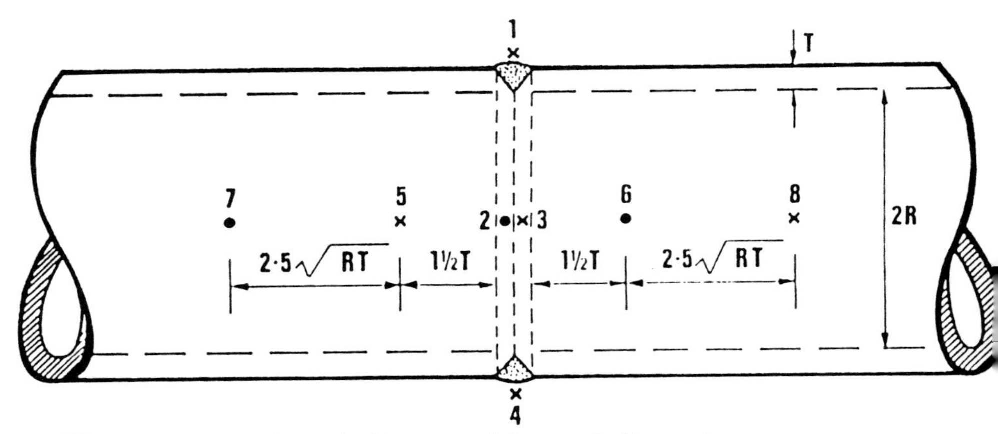

5.19 A thermocouple arrangement to ensure compliance with BS2633.

These stresses may be limited, therefore, by both achieving a sufficiently-wide temperature plateau about the weld and choosing a suitable band width. The temperature gradient away from the edge of the main heated band can be adjusted by insulating or by fitting additional heating circuits. British codes specify a minimum heated band width of $5\sqrt{Rt}$ (R = bore radius; t = wall thickness), whilst achieving a minimum temperature at the edges of this band of half peak temperature.

A method is needed to confirm full compliance with the requirements of BS2633 for high pressure pipework until repeatability is proven. It would be realistic, for a horizontal pipe weld between members of equal thickness, to devise an element/thermocouple arrangement as in Fig. 5.19. Where the parent members differ in thickness, account must be taken of the differing heat sinks each side of the weld caused by conduction.

The heating system should therefore either be displaced towards the heavier section (which is a complicated field instruction), or the system split to give side/side control as well as circumferential control. Such considerations would apply at pipe/flange joints, valve joints and T branch intersections. Furthermore, circumstances frequently arise where a butt weld undergoing a PWHT operation will generate unacceptably high thermal stresses in adjacent welds or areas of complexity. For example, stubs at T junctions in pipes should be of sufficient length that effects of the temperature gradient at the branch junction become insignificant. Some standards and codes of practice have taken this effect into account and propose practical means for control of the gradients and, if required, background heating on the main member.

This topic has been covered in some depth to emphasise the wide range of factors to be considered in producing a working procedure for what is apparently the simplest of localised heating operations. These factors are even more relevant in complex applications.

Chapter 6

Guidelines for inspection of heat treatment

Throughout the sequence of assembly of a joint in a steel or other metal fabrication and the subsequent welding, inspection points are clearly defined. They are the means by which assurance is given that the required standards of quality and workmanship have been achieved. Methods of inspection vary, from simple visual checks, through to technologically based techniques used in non-destructive testing and certain types of measurement. Irrespective of their basis, all these inspections can be measured against established criteria and documented to form a quality assurance record for each weld and cumulatively the complete and final assembly. This applies to all types of welded fabrication, be they structural work, pressure vessels or pipework systems.

The majority of welding carried out in a quality orientated environment is covered by weld procedures. These documents specify the joint geometry, which can be visually inspected and measured to verify compliance, and the welding technique and consumables, which can be verified by production control techniques and supervision by welding engineers.

On completion of welding, non-destructive examination is carried out to definitive procedures with reporting of defects to established criteria for acceptability, derived from national and sometimes company standards. Non-destructive (NDT) methods can produce an instantaneous visual representation, which is subsequently reported in an approved format – as with magnetic particle inspection, dye penetrant inspection and ultrasonics – or provide a permanent record – as with radiography.

The common factor in welding supervision and NDT is the need to train and certificate personnel to ensure levels of competency that underline the assurance of quality. Heat treatment is an optional operation associated with welding, and complete installations can be erected and welded without any heat treatment, but quality assurance to control weld quality and subsequent non-destructive examination is still required.

The need for heat treatment is dictated in most cases by the mandatory requirements of relevant national codes, for example ANSI B31, ASME VIII, BS5500, or by client specification with respect to service environment – low temperatures, caustic or sour service. Often in such circumstances information regarding the heat treatment process is limited to a specification of the heating cycle as part of the weld procedure. This information is usually an interpretation of the requirements of the relevant standard and to a lesser extent, client specification. Occasionally, a sketch showing thermocouple location may be included. Against such criteria meaningful inspection is limited, and usually a time/temperature chart is the only inspectable record of a heat treatment which can be included in the quality assurance documentation pack.

When nominating the welding and NDT specifications, consideration is given to the joint geometry and the immediately adjacent parent material. Often little reference is made to the overall component geometry. Normally any non-compliances associated with welding and NDT are confined within the body of the weld and its fusion faces. Remedial work is restricted to these locations and generally does not reflect on the overall integrity of the components or structure, unless there are metallurgical factors to consider.

However, where a post-weld heat treatment is involved, part or all of the structure joined by welding is heated to the requisite temperatures and therefore non-compliances affect the entire mass of metal at temperature, with potentially far-reaching consequences. Remedial work becomes more extensive and costly.

Essentials of good working practice

The basics of good furnace operation are well understood. Provided that a furnace is efficient, well insulated and exhibits good uniformity of temperature throughout the chamber, complete assemblies can be heated and cooled uniformly, in accordance with appropriate instructions, with assured repeatability.

The geometry and section thickness of the component to be treated must be reflected in the number and location of thermocouples and the choice of heat treatment cycle. Then assurance for the correct execution of the process rests with the skill of the operator, and correct functioning and calibration of the control and temperature recording equipment.

However, where local, in situ heat treatments are required the installed system operates on a one-off basis. It is therefore necessary to define the essentials of good working practice to enable meaningful inspections to be carried out and recorded. This gives a level of confidence that the installation is capable of performing the heat treatment to the betterment and not detriment of the structure. Also the records produced must correctly represent the specified cycle, temperature uniformity and, where required, control over thermal gradients.

The essentials of good working practice are:

1. *Accuracy of temperature recording instrument.* Calibration and maintenance must be carried out regularly by competent personnel using equipment which has been calibrated back to a traceable reference. To allow interpretation of the time/temperature chart, the chart speed must be known.
2. *Location of an appropriate number of thermocouples.* Locations must be chosen with respect to the size and geometry of the structure to provide a meaningful record of temperature profiles and gradients. As a basic requirement for butt welds and branch welds, recommended locations are shown in Fig. 6.1. However, a specific heat treatment procedure always indicates thermocouple locations.
3. *Installation of thermocouples by an approved method.* It has been proved that the most accurate method of attaching thermocouples to components to be heat treated is direct wire

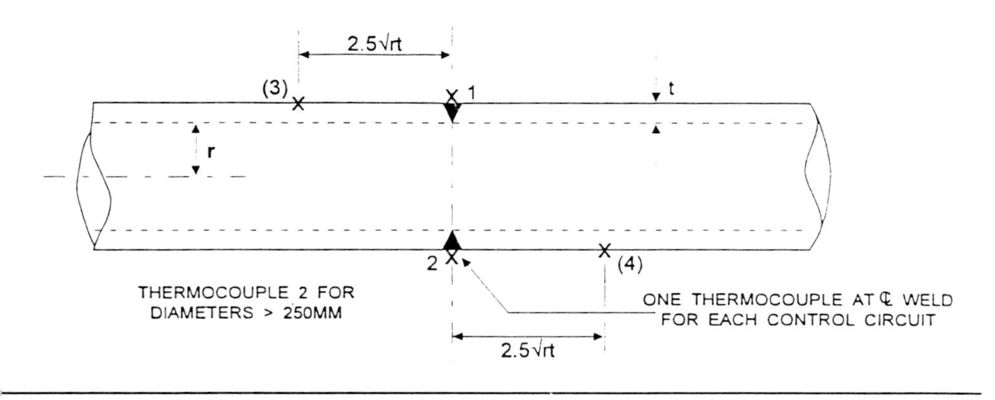

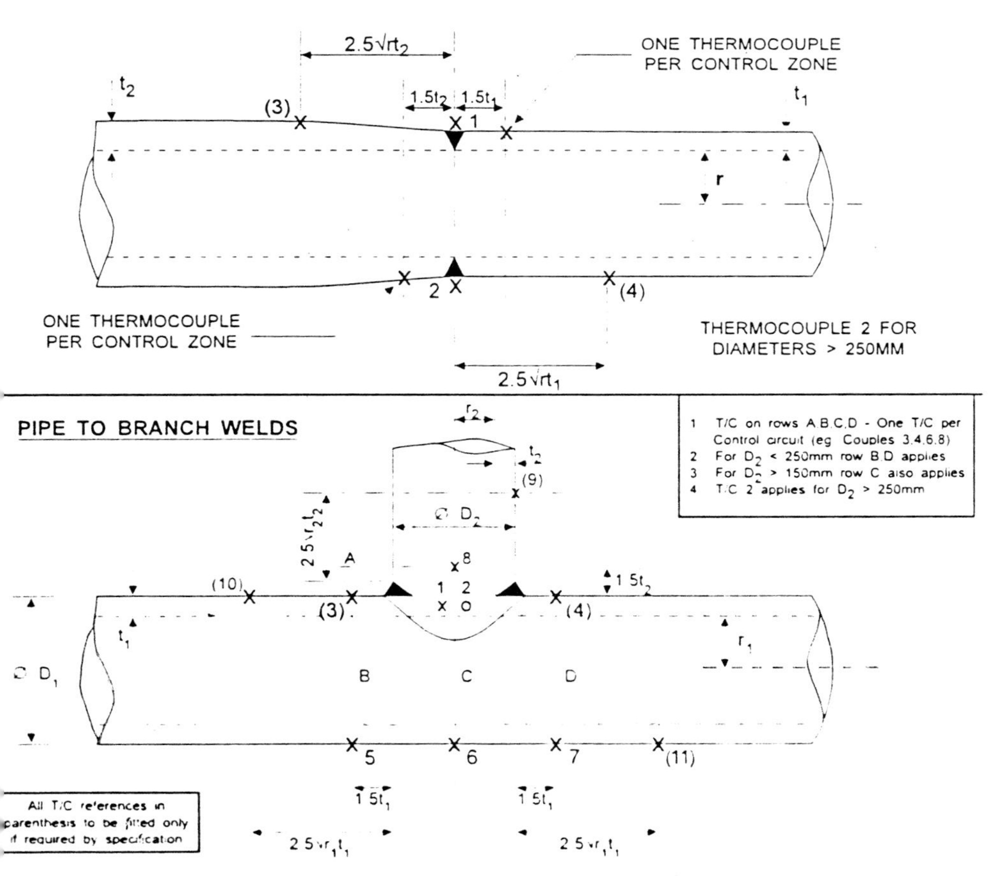

6.1 Typical thermocouple positions.

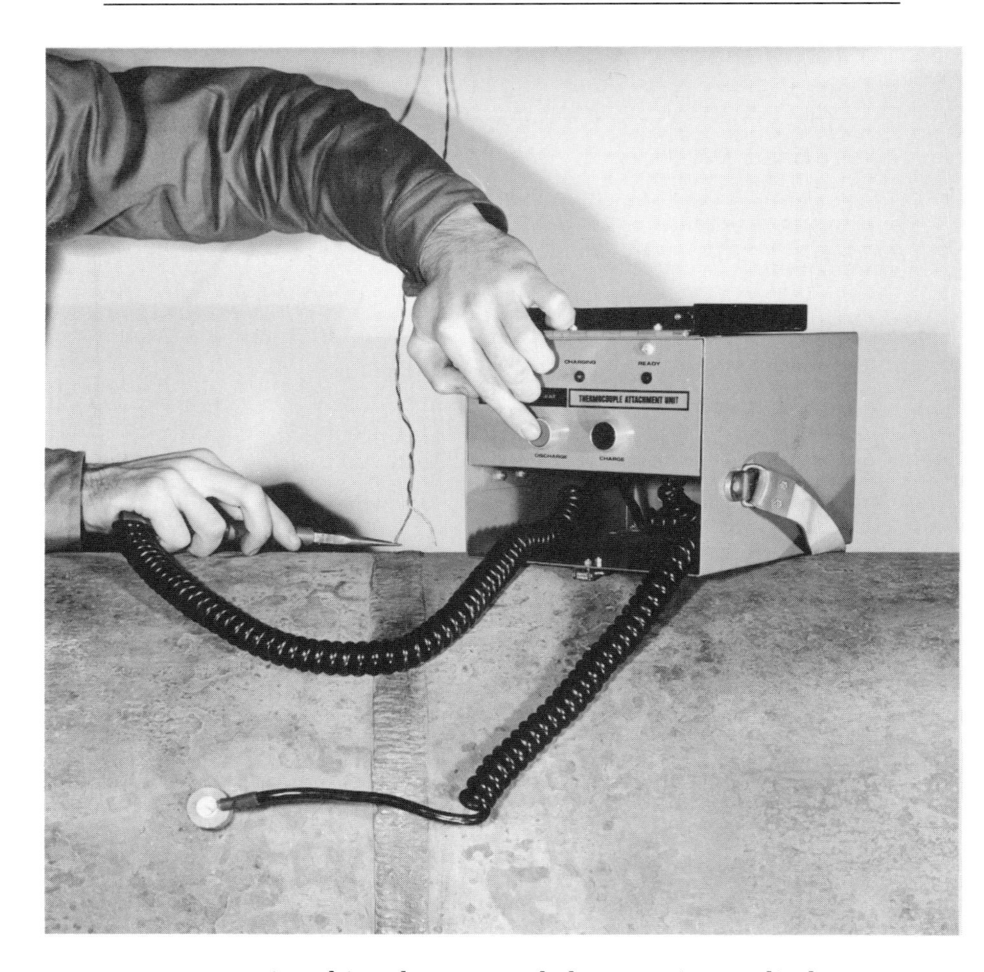

6.2 Attaching thermocouple by capacitance discharge (courtesy Cooperheat (UK) Ltd).

capacitance discharge. This technique, using portable battery operated equipment as shown in Fig. 6.2, attaches each wire of the thermocouple pair directly to the workpiece by resistance welding. Good welds have reasonable tensile strength, but poor shear strength. Therefore routeing is important so that no undue strain is placed on the hot junctions. When the gap between the two wires does not exceed 6 mm, the hot junction is actually on the surface where the temperature is being measured. There is no error in temperature reading, as is possible with slotted nut and other mechanical methods of attachment, provided that the thermocouple wire is correctly routed away from the hot zone and connected with the correct polarity to any compensating cable and the recording and/or controlling instruments. Any fault usually takes the form of a defective weld which manifests itself as an open circuit detection by the recording instrument. When the heat treatment has been completed, all thermocouple attachment points should be ground to remove residual local effects and checked by crack detection methods.

4. *Correct choice and location of heaters.* It is important to provide sufficient energy to raise the metal temperature and compensate for losses, over a width of hot zone capable of producing acceptable through wall temperature profiles. Because local heat treatments usually apply heat to only one surface of the component, normally the outer, the width of the heated band has to be sufficient to ensure that the temperature profile though the wall thickness subjects the complete weld and its heat affected zones to temperatures within the soak range for the appropriate time. The width of the heated zones has to reach certain minimum criteria. Comparison of codes shows variation in how the band width is specified, see Chapter 7.

   Considerable work has been carried out on this issue and the findings of Shifrin and Rich are notable in that they advise five times wall thickness.[28] A heater layout therefore requires the heated band width to be a function of wall thickness, and is usually specified as six or seven times this dimension to provide an additional margin.

   Flexible ceramic pad low voltage electrical resistance heaters have gained almost universal acceptance for on site in situ heat treatment. They are available in a range of standard sizes. Selection of the type and quantity of heaters will be designed to satisfy heat band width criteria whilst giving circumferential coverage without excessive gaps, which would give cold spots in the temperature profile.

   Overlapping of heaters is not advised as it can shorten their life. Experiments show that optimum circumferential gaps between heaters are 10 mm maximum for thickness up to 10 mm, and wall thickness or 30 mm maximum for thickness over 10 mm. Where longitudinal gaps between two or more rows of heaters are required the corresponding values are 10 mm, and wall thickness or 50 mm maximum. Longitudinal gaps can be included in the heated band width.

The elements selected should be held in close contact with the component surface to allow efficient heat transfer. Heaters on the undersides of horizontal structures should be examined for bulges. Heaters can be secured by stainless steel or mild steel bands or wires. Galvanized wires or other fixings likely to be detrimental to the welded structure must not be used. It is sometimes tempting to use the armouring from electric cables, which is soft and easily twisted, but more importantly usually found discarded in abundance on site and therefore *free*. This should not be used.

5. *Use of insulation to reduce heat losses.* The insulation applied to a local heat treatment serves two purposes:

(a) It reduces heat losses.
(b) It minimises temperature gradients away from the hot zone.

The material must be adequate with respect to thickness, density, thermal conductivity and temperature resistance. It must be applied to give total coverage of the required area without thinning or gaps or holes. As a rule, 40 mm (2×20 mm thickness) of a ceramic fibre having a minimum density of 96 kg/m^3 should cover the hot zone. Control of temperature gradients away from the hot zone can be achieved by using one layer (20 mm) of insulation to cover the required distance. Other insulation materials can be applied pro-rata to this requirement. Control of temperature gradients away from the hot zone is referenced in ASME VIII, BS5500 and BS2633, but not in ANSI, B31.1 or B31.3. A general guide is to use the formula $2.5\sqrt{Rt}$ to determine the minimum insulated distance beyond the hot zone, on both sides.

6. *Correct cable connection.* This applies to power cables and

thermocouples/compensating cables for accurate control. With flexible ceramic pads (FCP), the larger and more complex the component, the greater the number of individual discretely controlled heater zones required to control the heat treatment and uniformity of temperature profile. Therefore, neatly installed and clearly identified power cables, compensating cables, heater zones and thermocouples are required to ensure that the system is connected correctly. Typical faults are:

(a) Cross connection by thermocouple for one zone being incorrectly connected to the control circuit for another;
(b) Reverse polarity between thermocouple and compensating cables;
(c) Cross connection between programmer control leads and compensating cables and/or power cables.

Where there are two thermocouple functions, that is control/monitoring only, it can be worthwhile to differentiate between the types to ease identification when connecting.

7. *Selection of the appropriate heating cycle.* The cycle must comply with specified requirements. Where a heat treatment procedure has been prepared, either generic or specific, then criteria will be provided against which meaningful inspections can be carried out. These ensure adherence to the requirements. Where no detailed procedures are available, and welding engineers and/or inspectors are required to work from general statements of intent, or code interpretations, the above points of reference give a good basis for inspection.

Chapter 7

Manufacturing codes and standards

The majority of heat treatments carried out during fabrication of steel structures are determined by the requirements of the relevant Code or Standard for design and manufacture. Heat treatment may also be carried out when not mandatory for reasons of service environment or clients' instruction. There are circumstances which make it inadvisable, and alternative methods of achieving an acceptable end result have to be considered, but these solutions are not within the remit of this book.

Many countries have their own national codes and standards, and the requirements can differ widely. Within the scope of this book it is impossible to present a comprehensive overview, and comments are restricted to American codes, widely accepted throughout the petrochemical and process plant industries, and to the equivalent British Standards.

Information used in this chapter is extracted from the heat treatment sections in each document and relates to the editions current at the time of writing. It illustrates the basic requirements and reference must always be made to the particular Code or Standard for any particular application.

As the technology of welding progresses, the need for heat treatment is under constant review. For example, in BS5500 post-weld heat treatment is mandatory on carbon and carbon-manganese steel over 35 mm thick, whereas BS1500 and BS1515, both of which are now encompassed by the former had a threshold of 31.75 mm. BS2633 now requires preheat for carbon steels above

30 mm thick, whereas in 1966 the limit was 19 mm. Some welding development work has been carried out which identifies techniques for repair welding of thick carbon-manganese steel plates without post-weld heat treatment.[29] However, to waive an otherwise mandatory post-weld heat treatment requirement requires considerable justification, even for in situ repair, with carefully defined welding conditions supported by weld procedure qualification tests.

At the time of writing, further research under guidance of TWI (business name of The Welding Institute) and its American counterpart the Edison Welding Institute (EWI) was being carried out. This was reviewing welding without post-weld heat treatment and the need for circumferentially uniform heated bands in localised post-weld heat treatment of pressure vessels. This approach has greater significance for the repair and maintenance sectors of the petrochemical, process plant and power generation industries. Here in situ post-weld heat treatment often presents additional difficulties of location, restraint, and support.

All codes consider heat treatment requirements associated with steel structure fabrication in two categories:

- Preheat for welding – and in some instances thermal cutting;
- Post-weld heat treatment as a means of relieving stresses and achieving metallurgical change in the weld metal and heat affected zone.

Hydrogen release treatment before welding and post-heat for hydrogen diffusion after welding are not included, and are usually specified by the client as part of his weld procedure. Much information on these two topics is available from TWI.

In the codes, for preheat and post-weld heat treatment, materials are categorised based on alloy content, with temperatures and, where appropriate, hold times, specified for each group. Heating and cooling rates are generally universal and independent of steel type, although there are exceptions for specific steel grades and these are highlighted by sub classes or notes. Other information given in the codes relates to temperature differentials during heating and cooling, and determination of the minimum width of heated band required to perform localised post-weld heat

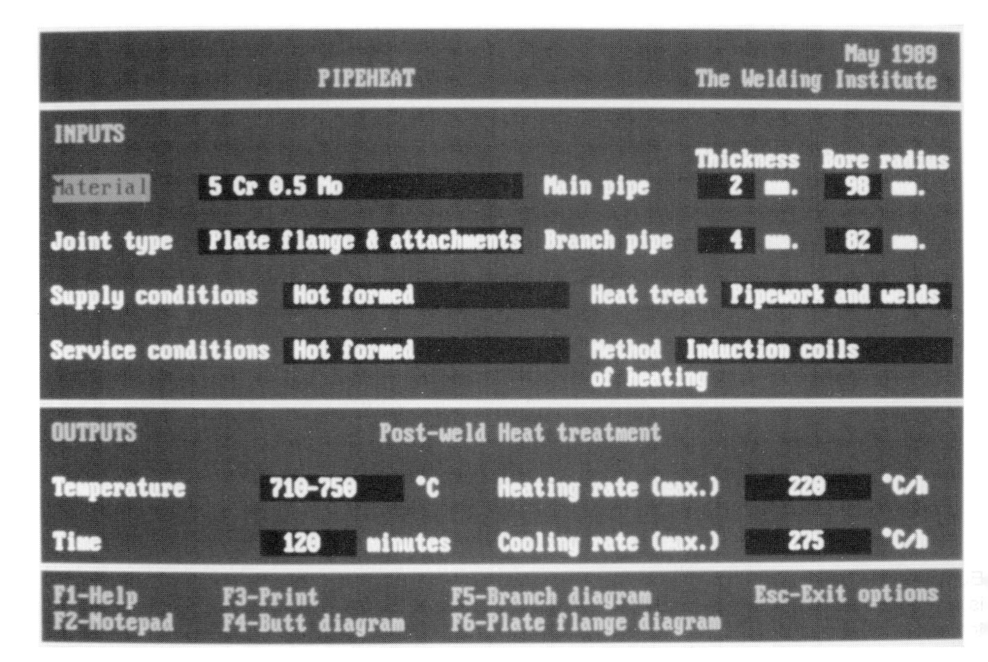

7.1 A post-weld heat treatment screen produced by PIPEHEAT software (courtesy TWI).

treatment, including in some cases comment on control of associated temperature gradients.

To illustrate the differing approaches between codes and standards, extracts from ASME VIII Division 1 and BS5500 covering construction of pressure vessels and ASME/ANSI B31.1, ASME/ANSI B31.3 and BS2633, covering pressure pipework systems, are presented in Tables 7.1 to 7.5. In the tables, any reference to joint thickness is that value defined by the relevant code. These tables clearly demonstrate the differences which must be understood by engineers involved with more than one code or standard.

For manufacture of pressure vessels, BS5500 requires mandatory preheat on all materials above a specified thickness, ASME VIII makes no mandatory requirement, but recommended values are given. Determination of post-weld heat treatment cycles shows

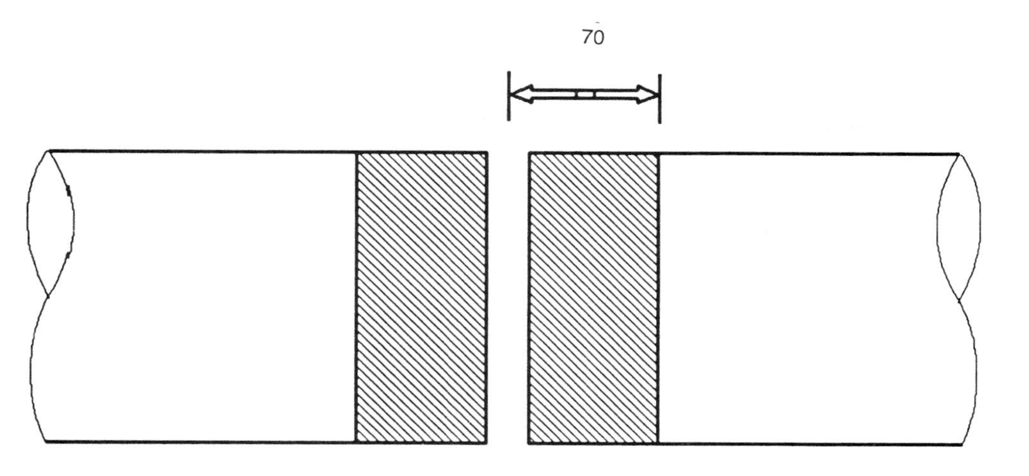

Butt joints shall be heat treated after the completion of all welding. Where local post-weld heat treatment is used, the temperature gradient shall be such that the length of material on either side of the weld at a temperature exceeding half the heat treatment temperature is at least:

$2.5\sqrt{rt} = 70$ Where r = bore radius
t = pipe thickness at weld
All dimensions in mm

7.2 Post-weld heat treatment requirements from PIPEHEAT software (courtesy TWI).

some small differences in evaluation of heating and cooling rates and soaking times and temperatures, but perhaps the largest difference is the way in which heat band widths for localised heat treatments are calculated. For a given size of vessel, BS5500 requires a larger heated band width than ASME VIII, the fundamental reason being that in the former, internal radius and thickness are taken into consideration, whilst in the latter only thickness is used. Additionally, BS5500 offers some comment on control of associated temperature gradients.

For fabrication of pressure pipework, BS2633 is more stringent than the ASME/ANSI counterparts. It is better defined, with more precise information on preheating temperatures, identifying root runs and filler runs separately. Post-weld heat treatment temperatures and requirements for localised post-weld heat treatment are compared with furnace heat treatment of complete assemblies.

It is significant to draw a comparison with ANSI B31.3, which does not specify any heating and cooling rates, but does identify measurement of hardness values as a method of verifying achievement of a satisfactory post-weld heat treatment. Also of interest are the relatively wide soak temperature ranges quoted in this standard.

Users of BS2633 can benefit from a microcomputer software programme PIPEHEAT, available from TWI. This programme offers the facility to produce heat treatment procedures for ferritic steel welds, butt joints, gusseted bends, branch connections, plate flanges and attachments. A typical VDU screen presentation is shown in Fig. 7.1. A sketch of the minimum width of heated band can also be produced (Fig. 7.2).

In selecting any preheat or post-weld heat treatment it is important to consider not only the code requirements, but also service performance. In some cases a knowledge of how the parent material was manufactured and of previous associated heat treatments is also essential. The adequacy of a series of heat treatments as part of the welding sequence should be established as an integral part of weld procedure qualification.

Table 7.1 Summary of BS5500 – Section 4.4

Material group	Preheat requirements for hydrogen controlled welding	PWHT requirements	Heating rate	Cooling rate	Requirements for localised PWHT
M0 Carbon M1 Carbon-manganese steels	t ≤ 30 mm . . . 5 °C min t > 30 mm . . . 100 °C min	Mandatory for thickness above 35 mm or 40 mm if impact properties above a specified level. 580–620 °C held for $2\frac{1}{2}$ minutes per mm of thickness (with minimum hold times)	240 °C/hr up to 25 mm thick or 6000 °C/hr divided by thickness (mm) Applies above 400 °C for ferritic steels	240 °C/hr up to 25 mm thick or 6000 °C/hr divided by thickness (mm) Applies above 400 °C for ferritic steels	1. For butt welds in vessels the heated band must extend at least 2.5√Rt each side of weld centre line where R = Internal Radius t = Shell Thickness 2. For branches and other attachments the heated band must extend at least 2.5√Rt from the edge of the connecting weld and cover a circumferential band around the entire vessel. 3. For all local heat treatments sufficient insulation is to be fitted to ensure that the weld & its heat affected zone achieve the required temperature & that the temperature at the edge of the heated band is not less than half the peak temperature. In addition the adjacent portions of the vessel outside the heated zone shall be thermally insulated such that the temperature gradient is not harmful. BS5500 recommends that the insulated band should extend 5√Rt each side of the weld centre line to achieve these requirements.
M2 C–Mo steels	t ≤ 12 mm . . . 20 °C min t > 12 mm . . . 100 °C min	Mandatory for thickness above 20 mm. 630–670 °C (minimum 60 minutes)			
M5 $3\frac{1}{2}$ Ni steels	To be specified dependent on plate thickness, welding consumables, welding process.	Requirements to be agreed. Where required the temperature range is 580–620 °C held for $2\frac{1}{2}$ min per mm of thickness (minimum 60 mins)			
M7 1 Cr $\frac{1}{2}$ Mo steels $1\frac{1}{2}$ Cr $\frac{1}{2}$ Mo steels	t ≤ 12 mm . . . 100 °C min t > 12 mm . . . 150 °C min	Mandatory for all thicknesses. 680–720 °C held for minimum 180 minutes			
M8 $\frac{1}{2}$ Cr $\frac{1}{2}$ Mo $\frac{1}{4}$V steels	t ≤ 12 mm . . . 100 °C min t > 12 mm . . . 200 °C min	Mandatory for all thicknesses. 680–720 °C held minimum 180 minutes			
M9 $2\frac{1}{4}$ Cr 1 Mo steel	t ≤ 12 mm . . . 150 °C min t > 12 mm . . . 200 °C min	Mandatory for all thicknesses. There are three temperature ranges 630–670 °C, 680–720 °C & 710–750 °C related to differing grades and final material properties. The minimum hold times also vary.			
M10 5 Cr $\frac{1}{2}$ Mo steel	All thicknesses 200 °C min	Mandatory for all thicknesses. 710–750 °C held minimum 120 minutes			

Table 7.2 Summary of ASME VIII Division 1 – Section UW40 and UCS-56

Material group	Preheat requirements	PWHT requirements	Heating rate	Cooling rate	Requirements for localised PWHT
P-No.1. Carbon steels Carbon-manganese steels	UW30 recommends that no welding is carried out at temperatures below 0 °F (–18 °C). Between 0 °F & 32 °F (0 °C) the surface of all areas within 3 in (75 mm) of the point where a weld is to be started should be heated to above 60 °F (16 °C) prior to welding.	Mandatory for thickness above $1\frac{1}{2}$ in (38 mm) [$1\frac{1}{4}$ in unless preheat used]. Minimum temperature 1100 °F (593 °C). Although the maximum differential is 150 °F, an upper limit of 1200 °F (649 °C) is often used. Hold times 1 hr/inch up to 2 in + 15 min per inch above 2 in.	400 °F (204 °C)/hr divided by thickness (inches) above 800 °F (426 °C) but no greater than 400 °F (204 °C)/hr Applies for all steel grades covered by Table UCS-56	500 °F (260 °C)/hr divided by thickness (inches) above 800 °F (426 °C) but no greater than 500 °F (200 °C)/hr Applies for all steel grades covered by Table UCS-56	1. For butt welds in vessels the heated band must extend at least two (2) times the shell thickness on each side of the greatest width of finished weld. 2. For pipe and tube butt welds the heated band must extend at least three (3) times the greatest width of finished weld on each side of the centre line. 3. For nozzles and other attachments the heated band must extend at least six (6) times the plate thickness beyond the connecting weld and cover a circumferential band around the entire vessel. 4. For all local heat treatments, the portions outside the heated band are to be protected so that the temperature gradients are not harmful.
P–No.3. C–$\frac{1}{2}$ Mo steels $\frac{1}{2}$ Cr–$\frac{1}{2}$ Mo steels	There is no mandatory requirement for preheat. Table UCS-56 lists preheat temperatures amongst conditions for relaxation of the mandatory requirements for PWHT. Other preheat requirements may be determined by clients' weld procedure qualification tests.	Mandatory over 5/8 in (16 mm) for Gr1 and 2 or all thicknesses in Gr3. Temperatures as for P1. Hold times as for P1.			
P-No.4. 1 Cr–$\frac{1}{2}$ Mo steels $1\frac{1}{4}$ Cr–$\frac{1}{2}$ Mo steels $2\frac{1}{2}$ Cr–$\frac{1}{2}$ Mo steels		Mandatory for thickness above 5/8 in (16 mm) OR all thicknesses dependent on actual grade. Temperatures as for P1. Hold times 1 hr/inch to 5 in + 15 min per inch above 5 in.			
P-No.5. $2\frac{1}{4}$ Cr–1 Mo 5 Cr–$\frac{1}{2}$ Mo 7 Cr–$\frac{1}{2}$ Mo 9 Cr–1 Mo		Mandatory under all conditions with some exceptions for pipework. Minimum temperatures 1250 °F (677 °C) or 1300 °F (704 °C). Hold times as for P4.			
P-No.9. $3\frac{1}{2}$ Ni steels		Mandatory for thicknesses above 5/8 in (16 mm). Temperatures as for P1 for P-9A 1100 °F (593 °C) to 1175 °F (640 °C) for P-9B. Hold times as for P4.			
P-No.10A		Mandatory for all thicknesses of SA487 CP1Q. Mandatory for thicknesses above 5/8 in (16 mm) all other materials. Temperatures as for P1. Hold times as for P4.			

Table 7.3 Summary of BS2633 – Sections 18 and 22

Material group	Preheat requirements	PWHT requirements	Heating rate	Cooling rate	Requirements for localised PWHT
Carbon & carbon-manganese steels ≤0.25 °C	t < 30 mm . . . 5 °C min t > 30 mm . . . 100 °C min	Mandatory for thicknesses above 35 mm & service temperatures below 0 °C. 580–620 °C held for $2\frac{1}{2}$ minutes per mm of thickness (minimum 30 minutes)	Up to 25 mm thick 220 °C/hr Over 25 mm thick 5500 °C/hr divided by thickness (mm) or 55 °C/hr whichever is greater Applies above 400 °C	Up to 25 mm thick 275 °C/hr Over 25 mm thick 6875 °C/hr divided by thickness (mm) or 55 °C/hr whichever is greater Applies above 400 °C	1. For butt welds the temperature gradient shall be such that the length of material on each side of the weld at a temperature exceeding half the heat treatment temperature is at least $2.5\sqrt{Rt}$ where R is the internal radius & t is the pipe thickness. 2. For branch welds the temperature gradient criterion above applies from the crotch of the weld for the main pipe & branch pipe using their respective dimensions.
Carbon & carbon-manganese steels 0.25 > C < 0.4	All thicknesses Root . . . 100 °C min Fill . . . 150 °C min	Mandatory for all thicknesses Temperature range 630–670 °C held as above			
C–Mo steels	t < 12.5 mm . . . 20 °C min t > 12.5 mm . . . 100 °C min	Mandatory for all thicknesses with exemptions. 630–670 °C held for minimum 60 minutes			
1 Cr–$\frac{1}{2}$ Mo $1\frac{1}{4}$ Cr–$\frac{1}{2}$ Mo steels	All thicknesses. Root . . . 100 °C min Fill . . . 100 °C min for t < 12.5 mm 150 °C min for t > 12.5 mm	Mandatory for all thicknesses with exemptions. 630–670 °C, minimum 120 minutes pipe & weld, 30 minutes weld only			

Table 7.3 (cont.)

Material group	Preheat requirements	PWHT requirements	Heating rate	Cooling rate	Requirements for localised PWHT
½ Cr–½ Mo–¼ V steel	All thicknesses. Root . . . 100 °C min Fill . . . 150 °C min t < 12.5 mm 200 °C min t > 12.5 mm	Mandatory for all thicknesses. Temperature range 680–720 °C held for a minimum of 180 minutes unless dia < 127 mm, t < 12.5 mm	For dia < 127 mm t < 12.5 mm 200 °C/hr otherwise 100 °C/hr or 6250 °C/hr divided by thickness (mm) whichever is lower. Applied above 400 °C.	For dia < 127 mm t < 12.5 mm 250 °C/hr otherwise 50 °C/hr	See previous page.
2¼ Cr–1 Mo steel	As for CMV	Mandatory for all thicknesses. Temperature range 680–720 °C held for 180 minutes unless dia 127 mm, t < 12.5 mm. For weld only minimum time is 60 minutes. An alternative soak range 710–750 °C is also quoted.			
5 Cr–½ Mo 7 Cr–½ Mo 9 Cr–1 Mo steels	All thicknesses Root . . . 150 °C min Fill . . . 200 °C min	Mandatory for all thicknesses. Temperature range 710–750 °C held for a minimum of 120 minutes.	Up to 25 mm thick 220 °C/hr Over 25 mm thick 5500 °C/hr divided by thickness (mm) or 55 °C/hr whichever is greater Applies above 400 °C	Up to 25 mm thick 275 °C/hr Over 25 mm thick 6875 °C/hr divided by thickness (mm) or 55 °C/hr whichever is greater Applies above 400 °C	
12 Cr–Mo–V(W) steels	All thicknesses Root . . . 150 °C min Fill . . . 150 to 300 °C Special precautions to be noted	Mandatory for all thicknesses. Temperature range 720–760 °C held for 180 minutes (pipe & weld) or 60 minutes minimum (weld only). Special note of concurrent preheat, post-heat, & post-weld heat treatment requirements.			
3½ Ni steels	All thicknesses Root . . . 100 °C min Fill . . . 150 °C min	580–620 °C when required, held for a minimum of 60 minutes.			

Table 7.4 Summary of ASME/ANSI B31-1. Chapter V – Sections 131 and 132

Material group	Preheat requirements	PWHT requirements	Heating rate	Cooling rate	Requirements for localised PWHT
P-No.1. C–Mn steels	50 °F (10 °C) minimum or 175 °F (80 °C) minimum if C > 0.30% and t > 1 in (25 mm)	Mandatory over $^3/_4$ in (19 mm). 1100 °F (600 °C) to 1200 °F (650 °C) held 1 hr/in minimum 15 minutes up to 2 in	600 °F (315 °C)/hr divided by $^1/_2$ thickness (in) above 600 °F but no greater than 600 °F/hr	As for heating	1. For butt welds the width of band heated to PWHT temperature must extend at least three (3) times the greatest wall thickness either side of the weld. 2. For nozzles and attachments the width of band heated to PWHT temperature must extend at least two (2) times the main pipe thickness beyond the connecting weld and cover a circumferential band around the pipe.
P-No.3. C–$^1/_2$ Mo $^1/_2$ Cr–$^1/_2$ Mo steels	50 °F (10 °C) minimum or 175 °F (80 °C) minimum if UTS > 60 ksi (413.7 MPa) or t > $^1/_2$ in (13 mm)	Mandatory over 5/8 in (16 mm). 1100 °F (600 °C) to 1200 °F (650 °C) held as for P-No.1.			
P-No.4. 1 Cr–$^1/_2$ Mo 1$^1/_4$ Cr–$^1/_2$ Mo 2$^1/_4$ Cr–$^1/_2$ Mo steels	50 °F (10 °C) minimum or 250 °F (120 °C) minimum if UTS > 60 ksi (413.7 MPa) or t > $^1/_2$ in (13 mm)	Mandatory except for smaller pipes under certain conditions 1300 °F (700 °C) to 1375 °F (750 °C) held as for P-No.1.			

Table 7.4 (cont.)

Material group	Preheat requirements	PWHT requirements	Heating rate	Cooling rate	Requirements for localised PWHT
P-No.5. $2\frac{1}{4}$ Cr–1 Mo 5 Cr–$\frac{1}{2}$ Mo 7 Cr–$\frac{1}{2}$ Mo 9 Cr–1 Mo	300 °F (150 °C) minimum or 400 °F (200 °C) minimum if UTS > 60 ksi (4137 MPa) or Cr > 6.0% and t > $\frac{1}{2}$ in (13 mm)	Mandatory unless all of following are met: OD < NPS4; t < $\frac{1}{2}$ in (13 mm); Cr < 3.0%; C < 0.15%; Preheat 300 °F. 1300 °F (700 °C) to 1400 °F (760 °C) held as for P1.		See previous page.	
P-No.9. $3\frac{1}{2}$ Ni steels	250 °F (120 °C) minimum for P-No.9A. 300 °F (150 °C) minimum for P-No.9B.	P-No.9A – as for P-No.1. P-No.9B – above 5/8 in (16 mm) 1100 °F (600 °C) to 1175 °F (630 °C) held as for P-No.1.			

Table 7.5 Summary of ASME/ANSI B31.3. Chapter IX, Part 9, Sections 330 and 331

Material group	Preheat requirements	PWHT requirements	Heating rate	Cooling rate	Requirements for localised PWHT
	GENERAL REQUIREMENTS 1. Preheat zone to extend 1 in (25 mm) beyond edge of weld. 2. Preheat requirements apply if ambient temperature falls below 32 °F (0 °C)				
P-No.1. Carbon steel	Recommended 50 °F (10 °C) minimum or 175 °F (79 °C) minimum if UTS > 71 ksi (490 MPa) or t > 1 in (25 mm)	Mandatory over ¾ in (19 mm). 1100 °F (593 °C) to 1200 °F (649 °C) held 1 hr/in minimum or 1 hr.	Not specified	Not specified	A circumferential band of the pipe, and branch where applicable to be heated so that the required PWHT temperature exists over the entire pipe section(s), gradually diminishing beyond the ends of the band which includes the weldment and at least 1 in (25 mm) beyond the edges thereof.
P-No.3. Alloy steels Cr < ½%	Recommended 50 °F (10 °C) minimum or 175 °F (79 °C) minimum if UTS > 71 ksi (490 MPa) or t > ½ in (12.7 mm)	Mandatory over ¾ in (19 mm) unless UTS ≤ 71 ksi 1100 °F (593 °C) to 1325 °F (718 °C) held 1 hr/in minimum 1 hr. Maximum hardness 225 Brinell			

Table 7.5 (cont.)

Material group	Preheat requirements	PWHT requirements	Heating rate	Cooling rate	Requirements for localised PWHT
P-No.4. Alloy steels $\frac{1}{2} < Cr \leq 2$	Required for all thicknesses 300 °F (149 °C) minimum	Mandatory over $\frac{1}{2}$ in (12.7 mm) unless UTS ≤ 71 ksi. 1300 °F (704 °C) to 1375 °F (746 °C) held 1 hr/in minimum 2 hr. Maximum hardness 225 Brinell.		See previous page.	
P-No.5. Alloy steels $2\frac{1}{4} \leq Cr \leq 10$ (≤ 3 Cr ≤ 0.15C) (>3 Cr or >0.15C)	Required for all thicknesses 350 °F (177 °C) minimum	Mandatory over $\frac{1}{2}$ in (12.7 mm). 1300 °F (704 °C) to 1400 °F (760 °C) held 1 hr/in minimum 2 hr. Maximum hardness 241 Brinell.			
P-No.9. A & B nickel alloy steels	Recommended for all thicknesses 200 °F (93 °C) minimum	Mandatory above $\frac{3}{4}$ in (19 mm). 1100 °F (593 °C) to 1175 °F (635 °C) held $\frac{1}{2}$ hr per in, minimum 1 hr.			

References

1 Bailey N, *Weldability of Ferritic Steels,* Abington Publishing, 1994.
2 Granville B A, Baker R G and Watkinson F, *British Weld J* 14(6), 1967, pp 337–342.
3 Cottrell C L M, *Welding and Metal Fabrication* April 1990, pp 178–183.
4 Bailey N, Coe F R, Gooch T G, Hart P H M, Jenkins N, Pargeter R J, *Welding Steels without Hydrogen Cracking*, Second Edition, Abington Publishing, 1993.
5 Matharu I S and Hart P H M, TWI Res Report 290/1985.
6 Rodwell M H and Evans S M, TWI Res Report, 296/1986.
7 Winterton K, *Weld Res Supp*, 26(6), 1961, pp 253s–258s.
8 Still J R, *Welding and Metal Fabrication* Aug/Sept 1994, pp 331–339.
9 Evans U R, *The Corrosion and Oxidation of Metals*, Edward Arnold (Publishers) Ltd, London 1960 & 1968.
10 Gurney T R, *British Weld J*, 7(6), 1960, pp 415–431.
11 Burdekin F M, *Weld J, Res Supp*, 47(1), 1968, pp 129s–139s.
12 BS5500:1991.
13 Scifo A, *Welding International* 4(9), 1990, pp 703–713.
14 Bentley K P, *British Weld J*, 11(10), 1964, pp 507–515.
15 Murray J D, *British Weld J*, 14(8), 1967, pp 447–456.
16 American Society of Metals, Conference on Temper Embrittlement, December 1967.
17 Cole C L and Jones J D, *Stainless Steels*, The Iron & Steel Institute, London, 1969, pp 71–76.

18 Younger R N, Haddrill D M and Baker R G, *J Iron and Steel Inst*, 201, 1963, pp 693–698.

19 Samans C H, *Corrosion*, 20(8), 1964, pp 256t–262t.

20 Malone M O, *Weld Res Supp*, 32(6), 1967, pp 241s–253s.

21 Charles J, *Duplex Stainless Steels,* Les editions de physique, 1991 Volume 1, pp 3–48.

22 Dilthey V, Grobecker J, Rao K P and Sabat H K, *Welding and Cutting*, 9, 1994, pp E155–E159.

23 Woodley C C, Burdekin F M and Wells A A, *British Weld J*, 13(3), 1966, pp 165–173.

24 Woodley C C and Burdekin F M, *British Weld J*, 13(6), 1966, pp 387–397.

25 Egan G R, *WI Res Bulletin* 9 1968, pp 231–234.

26 Guan Q, Guo D L, Li C Q, Leggatt R H, *Welding in the World*, 33 (3), 1994, pp 160–167.

27 Cottrell D J, *Electric Resistance and Gas Fired Techniques for Post Weld Heat Treatment Heat Treatment of Metals*, 1989, pp 104–110.

28 Shifrin E G and Rich M I, *Weld Journal*, 52(12), 1973, pp 792–799.

29 Jones R L, TWI Res Report 335/1987.

Index